# Event-Driven Architecture in Golang

Building complex systems with asynchronicity and
eventual consistency

**Michael Stack**

BIRMINGHAM—MUMBAI

# Event-Driven Architecture in Golang

**Group Product Manager**: Gebin George
**Publishing Product Manager**: Gebin George
**Senior Editor**: Rounak Kulkarni
**Technical Editor**: Pradeep Sahu
**Copy Editor**: Safis Editing
**Project Coordinator**: Manisha Singh
**Proofreader**: Safis Editing
**Indexer**: Hemangini Bari
**Production Designer**: Prashant Ghare
**Developer Relations Marketing Executive**: Sonakshi Bubbar
**Business Development Executive**: Bhanu Rangani

First published: November 2022
Production reference: 1281022

Published by Packt Publishing Ltd.
Livery Place
35 Livery Street
Birmingham
B3 2PB, UK.

ISBN 978-1-80323-801-2

www.packt.com

*To my father, William Stack, for introducing me to the world of computers and instilling in me a passion for technology. To my wife, Kate, for enduring my long hours and weekends locked away working on the book. To my son, Malcolm, most of all, since I doubt I could have finished without his unceasing support, drawings, messages, and words of encouragement.*

*– Michael Stack*

# Contributors

## About the author

**Michael Stack** is a solutions architect who has over 20 years of expertise in the software development industry. Throughout the course of his career, he has developed a variety of applications, including company blogs and intranets, festival ticketing systems, multiplayer games, and national park management software. Currently, his principal focus is on the development of microservices and other distributed applications. He is passionate about using Go and has been doing so for the better part of a decade for both personal and professional projects.

*I would like to thank everyone who supported and encouraged me, especially my wife, Kate, my father, my son, and Packt.*

# About the reviewers

**Dimas Yudha Prawira** is a father, engineer, public speaker, tech community leader, and tech organizer. He has 18 years of experience with software engineering and has worked for various companies, including PT Telekomunikasi Indonesia (Telkom), Kudo x Grab, OVO, and RCTI+. He loves the Go programming language, Java, and talking about code. He is most interested in microservices, SOA, APIs, software architecture, embedded systems, and more.

When he is not working, he uses his time to read books, watch movies, or just play with his family.

**Samantha Coyle** is a Go backend engineer, speaker, and mentor with a love for all things Go and open source. She spends her days developing Go microservices, bringing in new features, observability, improved testing, and best practices. She is a software engineer for the **Internet of Things** (**IoT**) group at Intel, where she enables healthcare solutions using Go and Open Source technology at the edge and has worked on computer vision-based smart city solutions and industrial applications. Samantha explores thought leadership avenues, including reviewing Go textbooks, speaking at GopherCon, Grace Hopper Conference, and Open Source Summit in 2022, attaining her CKAD certification, and volunteering to mentor early career professionals to grow as Go engineers.

# Table of Contents

3

## Design and Planning                             43

# Part 2: Components of Event-Driven Architecture

4

## Event Foundations                               75

# 5

# Tracking Changes with Event Sourcing        99

# 6

# Asynchronous Connections        141

# Part 3: Production Ready

## 10

## Testing                                                    259

## 11

## Deploying Applications to The Cloud                        297

## 12

## Monitoring and Observability                                                      325

## Index                                                                                        349

## Other Books You May Enjoy                                                        352

# Preface

Companies are adopting **event-driven architecture** (**EDA**) as their web applications grow in size and complexity. Applications that communicate using events are easier to develop and scale. Adding or developing your application around real-time interactions becomes easier with EDA.

Direct point-to-point communication between microservices inevitably leads to the development of a distributed monolith, which is just a monolith with extra and unnecessary complexity. EDA is an architecture that helps organizations to decouple microservices and avoid developing another distributed monolith.

Choosing a new architecture for your next application or deciding to refactor an existing one can be fraught with known and unknown challenges. It is my intention and this book's goal to provide you with enough examples and knowledge to give you a great head start should you decide to take the development of an EDA.

In this book, we will discuss and cover EDA concepts and related topics with the help of a small modular monolith demonstration application. We will use this application to take a journey through the concepts and topics to convert the synchronous mechanisms used by the application into asynchronous communication mechanisms.

## Who this book is for

This architecture book is for developers working with microservices, or those architecting and designing new applications that will be built with microservices. Intermediate-level knowledge of Go is required to make the most of the examples and concepts in this book. Developers with a background in any programming language and experience working with microservices should still find the concepts and explanations useful.

## What this book covers

*Chapter 1*, *Introduction to Event-Driven Architectures*, introduces EDA.

*Chapter 2*, *Supporting Patterns in Brief*, covers helpful patterns such as domain-driven design, domain-centric architectures, and application architectures.

*Chapter 3*, *Design and Planning*, explores the ways to discover the capabilities and features of an application using EventStorming and other methods.

*Chapter 4, Event Foundations*, introduces the Mallbots modular monolith application and domain events.

*Chapter 5, Tracking Changes with Event Sourcing*, introduces event sourcing and leads you through the development of event-sourced aggregates.

*Chapter 6, Asynchronous Connections*, covers adding asynchronous communication using event messages.

*Chapter 7, Event-Carried State Transfer*, expands on the use of message-based communication between components.

*Chapter 8, Message Workflows*, covers the concept of distributed transactions and introduces orchestrated sagas.

*Chapter 9, Transactional Messaging*, explores the use of message inboxes and outboxes to reduce data loss.

*Chapter 10, Testing*, discusses the concept of a testing strategy and leads you through testing an event-driven application.

*Chapter 11, Deploying Applications to the Cloud*, covers the use of infrastructure as code and deploying an application as microservices.

*Chapter 12, Monitoring and Observability*, discusses how to monitor a distributed application and make it observable with logging, metrics, and distributed tracing.

## To get the most out of this book

This book is written with the expectation that you can execute the demonstration application to understand and view the code changes that have been made in each chapter as the application is refactored. To get the most out of the book, it is recommended you read the chapters in order, as the chapters will reference code that has been modified in the previous chapter.

| Software/hardware covered in the book | Operating system requirements |
|---|---|
| Go 1.18+ | Windows, macOS, or Linux |
| Docker 20.10.x | Windows, macOS, or Linux |
| NATS 2.4 | Windows, macOS, or Linux |

Most of the development for this book was done in Windows 10, but the code was tested to run in **Windows Subsystem for Linux 2 (WSL 2)** in Ubuntu 20.04 and tested to run on a Mac. You are expected to run the application and its dependencies within a Docker compose environment. Instructions to use Docker are given wherever possible to minimize installing new software on your machine.

**If you are using the digital version of this book, we advise you to type the code yourself or access the code from the book's GitHub repository (a link is available in the next section). Doing so will help you avoid any potential errors related to the copying and pasting of code.**

You can follow the author on GitHub (`https://github.com/stackus`) or make a connection with them on LinkedIn (`https://www.linkedin.com/in/stackmichael`).

# Download the example code files

You can download the example code files for this book from GitHub at `https://github.com/PacktPublishing/Event-Driven-Architecture-in-Golang`. If there's an update to the code, it will be updated in the GitHub repository.

We also have other code bundles from our rich catalog of books and videos available at `https://github.com/PacktPublishing/`. Check them out!

# Download the color images

We also provide a PDF file that has color images of the screenshots and diagrams used in this book. You can download it here: `https://packt.link/qgf10`.

# Conventions used

There are a number of text conventions used throughout this book.

`Code in text`: Indicates code words in text, database table names, folder names, filenames, file extensions, pathnames, dummy URLs, user input, and Twitter handles. Here is an example: "If all the participants have responded positively, then the coordinator will send a `COMMIT` message to all of the participants and the distributed transaction will be complete."

A block of code is set as follows:

```
BEGIN;
-- execute queries, updates, inserts, deletes …
PREPARE TRANSACTION 'bfa1c57a-d99d-4d74-87a9-3aaabcc754ee';
```

When we wish to draw your attention to a particular part of a code block, the relevant lines or items are set in bold:

```
func NewCommandHandlers(
    app application.App,) ddd.CommandHandler
        [ddd.Command] {
    return commandHandlers{
        app: app,
```

Any command-line input or output is written as follows:

```
--- PASS: TestApplication_AddItem (0.00s)
    --- PASS: TestApplication_AddItem/NoBasket (0.00s)
    --- PASS: TestApplication_AddItem/NoProduct (0.00s)
    --- PASS: TestApplication_AddItem/NoStore (0.00s)
    --- PASS: TestApplication_AddItem/SaveFailed (0.00s)
    --- PASS: TestApplication_AddItem/Success (0.00s)
PASS
```

**Bold**: Indicates a new term, an important word, or words that you see on screen. For instance, words in menus or dialog boxes appear in **bold**. Here is an example: "The **Customers** module remains uncoupled from the **Order Processing** module because we do not have any explicit ties to the **Order Processing** module in this handler."

> **Tips or Important Notes**
> Appear like this.

# Get in touch

Feedback from our readers is always welcome.

**General feedback**: If you have questions about any aspect of this book, email us at customercare@packtpub.com and mention the book title in the subject of your message.

**Errata**: Although we have taken every care to ensure the accuracy of our content, mistakes do happen. If you have found a mistake in this book, we would be grateful if you would report this to us. Please visit www.packtpub.com/support/errata and fill in the form.

**Piracy**: If you come across any illegal copies of our works in any form on the internet, we would be grateful if you would provide us with the location address or website name. Please contact us at copyright@packt.com with a link to the material.

**If you are interested in becoming an author**: If there is a topic that you have expertise in and you are interested in either writing or contributing to a book, please visit authors.packtpub.com.

## Share Your Thoughts

Once you've read *Event-Driven Architecture in Golang*, we'd love to hear your thoughts! Scan the QR code below to go straight to the Amazon review page for this book and share your feedback.

https://packt.link/r/1803238011

Your review is important to us and the tech community and will help us make sure we're delivering excellent quality content.

# Download a free PDF copy of this book

Thanks for purchasing this book!

Do you like to read on the go but are unable to carry your print books everywhere? Is your eBook purchase not compatible with the device of your choice?

Don't worry, now with every Packt book you get a DRM-free PDF version of that book at no cost.

Read anywhere, any place, on any device. Search, copy, and paste code from your favorite technical books directly into your application.

The perks don't stop there, you can get exclusive access to discounts, newsletters, and great free content in your inbox daily

Follow these simple steps to get the benefits:

1.  Scan the QR code or visit the link below

https://packt.link/free-ebook/9781803238012

2.  Submit your proof of purchase
3.  That's it! We'll send your free PDF and other benefits to your email directly

# Part 1: Event-Driven Fundamentals

This first part will provide an understanding of what **event-driven architecture** (**EDA**) is and the benefits of using it for your next application. We will also be introduced to the application we will be working with. This part will also cover some helpful patterns that can be helpful for the adoption and development of EDA. Then, it will introduce methods for planning an application using EventStorming.

This part consists of the following chapters:

- *Chapter 1, Introduction to Event-Driven Architecture*
- *Chapter 2, Supporting Patterns in Brief*
- *Chapter 3, Design and Planning*

# 1
# Introduction to Event-Driven Architectures

**Event-driven architecture** (EDA) is the foundational design of an application's communication of state changes around an asynchronous exchange of messages called **events**. The architecture allows applications to be developed as a highly distributed and loosely coupled organization of components. Probably predominantly, the most well-known arrangement of components today is the microservices architecture for applications.

Our world is made up of events—they're happening everywhere around us. A simple act of waking up in the morning becomes an event the instant it occurs. The same goes for the act of purchasing a book. Whether or not it was recorded that way in some database, somewhere, it was considered an event. Since it has occurred, several other operations might have sprung from it.

Just as companies looked at microservices a decade ago to address issues such as *web-scale*, EDA is gaining in interest and proponents of continuing that journey to help with *global-scale*.

It is my goal in this chapter to introduce you to the concepts and components of EDA and its applications that we will be using to demonstrate what EDA has to offer. We'll also be taking a grounded look at the benefits and reasons to use EDA and the challenges you're likely to encounter when starting a new greenfield project or adding select concepts and components to an existing project.

Whether you're looking to start a new project with an event-driven approach or looking to break up a monolithic application into modules or further into microservices, this book will give you the information and patterns necessary to implement EDA where you need it.

In this chapter, we're going to cover the following main topics:

- An exchange of facts
- The MallBots application
- Benefits of EDA
- Challenges of EDA

# Technical requirements

We will be developing using Go and using Docker to run our application within containers. Visit the following to locate installers for your operating system:

- Go installers can be found at https://go.dev/doc/install
- Docker installers can be found at https://docs.docker.com/desktop/

Go 1.17 or higher is required to run the code from this book.

# An exchange of facts

Three different uses or patterns exist that can be called EDA individually or altogether, as follows:

- Event notifications
- Event-carried state transfer
- Event sourcing

In this book, we will be covering each of these patterns, going over their uses and both when to use them and when you might not.

## Event notifications

Events can be used to notify something has occurred within your application. A notification event typically carries the absolute minimum state, perhaps even just the **identifier** (**ID**) of an entity or the exact time of the occurrence of their payload. Components that are notified of these events may take any action they deem necessary. Events might be recorded locally for auditing purposes, or the component may make calls back to the originating component to fetch additional relevant information about the event.

Let's see an example of `PaymentReceived` as an event notification in Go, as follows:

```
type PaymentReceived struct {
    PaymentID string
}
```

Here is how that notification might be used:

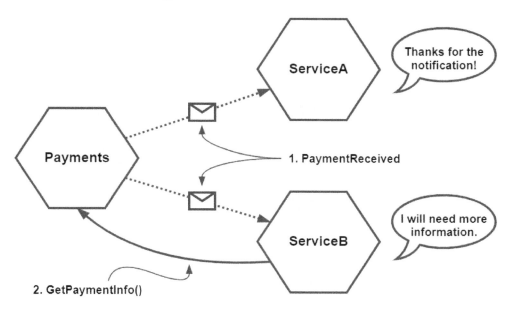

Figure 1.1 – PaymentReceived as an event notification

*Figure 1.1* shows the `PaymentReceived` notification being received by two different services. While **ServiceA** only needed to be notified of the event, **ServiceB** will require additional information and must make a call back to the **Payments** service to fetch it.

## Event-carried state transfer

Event-carried state transfer is an asynchronous cousin to **representational state transfer** (**REST**). In contrast with REST's on-demand pull model, event-carried state transfer is a push model where data changes are sent out to be consumed by any components that might be interested. The components may create their own local cached copies, negating any need to query the originating component to fetch any information to complete their work.

Let's see an example of `PaymentReceived` as an event-carried state transfer, as follows:

```
type PaymentReceived struct {
    PaymentID    string
    CustomerID   string
    OrderID      string
    Amount       int
}
```

In this example for event-carried state transfer, we've included some additional IDs and an amount collected, but more detail could be added to provide as much detail as possible, as illustrated in the following diagram:

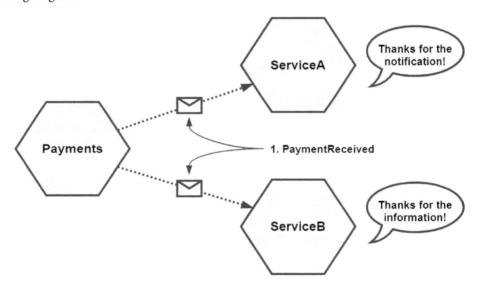

Figure 1.2 – PaymentReceived as an event-carried state change

When the `PaymentReceived` event is sent with additional information, it changes how downstream services might react to it. We can see in *Figure 1.2* that **ServiceB** no longer needs to call the **Payments** service because the event it has received already contains everything it requires.

## Event sourcing

Instead of capturing changes as irreversible modifications to a single record, those changes are stored as events. These changes or streams of events can be read and processed to recreate the final state of an entity when it is needed again.

When we use event sourcing, we store the events in an **event store** rather than communicating them with other services, as illustrated in the following diagram:

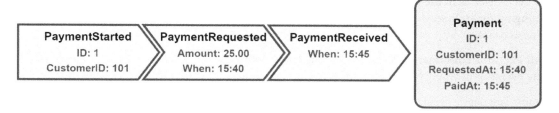

Figure 1.3 – Payment data recorded using event sourcing

In *Figure 1.3*, we see the entire history of our data is kept as individual entries in the event store. When we need to work with a payment in the application, we would read all the entries associated with that record and then perform a left fold of the entries to recreate the final state.

## Core components

You will observe that four components are found at the center of all event patterns, as illustrated in the following diagram:

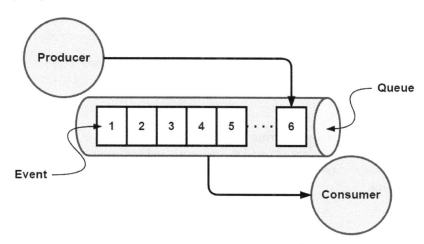

Figure 1.4 – Event, queue, producer, and consumer

## *Event*

At the heart of EDA is the **event**. In EDA terms, it is an occurrence that has happened in the application. The event itself is in the past and it is an immutable fact. Some examples of events are customers signing up for your services, payments being received for orders, or failed authentication attempts for an account.

With EDA, the consumers of these events may know nothing about what caused the production of these events or have any relationship or connection with them, but only with the event itself.

In most languages, events are simple value objects that contain state. An event is equal to another if all the attributes are the same. In Go, we would represent an event with a simple struct, such as this one for `PaymentReceived`:

```
type PaymentReceived struct {
    PaymentID string
    OrderID   string
    Amount    int
}
```

Events should carry enough data to be useful in capturing the change in the application state that they're meant to communicate. In the preceding example, we might expect that this event is associated with some payment, and the specific payment is identified by the queue name or as some metadata passed along with the event instead of the `PaymentID` field in the body of the event being necessary.

The amount of information required to include in an event's payload matters to all events, the event notification, the event-carried state transfer, and for the changes recorded with event sourcing.

## *Queues*

**Queues** are referred to by a variety of terms, including bus, channel, stream, topic, and others. The exact term given to a queue will depend on its use, purpose, and sometimes vendor. Because events are frequently—but not always—organized in a **first-in, first-out** (**FIFO**) fashion, I will refer to this component as a queue.

### Message queues

The defining characteristic of a **message queue** is its lack of event retention. All events put into a message queue have a limited lifetime. After the events have been consumed or have expired, they are discarded.

You can see an example of a message queue in the following diagram:

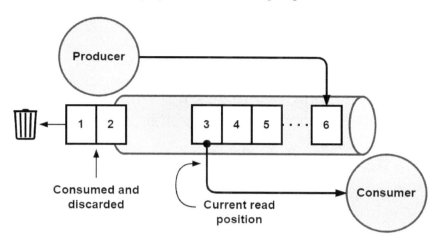

Figure 1.5 – Message queue

A message queue is useful for simple **publisher/subscriber (pub/sub)** scenarios when the subscribers are actively running or can retrieve the events quickly enough.

### Event streams

When you add event retention to a message queue, you get an **event stream**. This means consumers may now read event streams starting with the earliest event, from a point in the stream representing their last read position, or they can begin consuming new events as they are added. Unlike message queues, which will eventually return to their default empty state, an event stream will continue to grow indefinitely until events are removed by outside forces, such as being configured with a maximum stream length or archived based on their age.

The following diagram provides an example of an event stream:

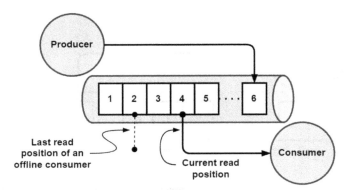

Figure 1.6 – Event stream

When you need retention and the ability to replay events, an event stream should be used instead of a message queue.

**Event stores**

As the name implies, an **event store** is an append-only repository for events. Potentially millions of individual event streams will exist within an event store. Event stores provide **optimistic concurrency** controls to ensure that each event stream maintains strong consistency. In contrast to the last two queue examples, an event store is typically not used for message communication.

You can see an example of an event store in the following screenshot:

| Events Table | | | | | |
|---|---|---|---|---|---|
| ID | ObjectID | ObjectType | Version | Type | Data |
| 101 | 1 | Payment | 1 | PaymentStarted | { ... } |
| 102 | 1 | Payment | 2 | PaymentRequested | { ... } |
| 103 | 1 | Payment | 3 | PaymentReceived | { ... } |
| 104 | 2 | Payment | 1 | PaymentStarted | { ... } |
| ... | ... | ... | ... | ... | ... |

Events for **Payment**: 1
Events for **Payment**: 2

Figure 1.7 – Event store

Event stores are used in conjunction with event sourcing to track changes to entities. The top three rows of *Figure 1.7* depict the event-sourcing example events from *Figure 1.3*.

### Producers

When some state in the application has changed, the producer will publish an event representing the change into the appropriate queue. The producer may include additional metadata along with the event that is useful for tracking, performance, or monitoring. The **producers** of the events will publish it without knowing what the consumers might be listening to. It is essentially a fire-and-forget operation.

### Consumers

**Consumers** subscribe to and read events from queues. Consumers can be organized into groups to share the load or be individuals reading all events as they are published. Consumers reading from streams may choose to read from the beginning of a stream, read new events from the time they started listening, or use a cursor to pick up from where they left the stream.

## Wrap-up

Equipped with the types of events we will be using and the knowledge of the components of the patterns involved, let's now look at how we'll be using them to build our application.

# The MallBots application

We're going to be building a small application that simulates a retail experience coupled with some futuristic shopping robots. We will be building the backend services that power this application. A high-level view of the components involved is shown here:

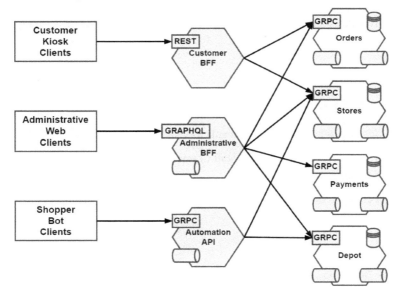

Figure 1.8 – High-level view of the MallBots application components

## The pitch

> *"We have developed incredible robots to save the time of people shopping at the mall. Customers will now have access to a kiosk that would facilitate the selection of items from available stores that customers do not wish to visit. After completing their selections, the customer is free to do other shopping or directly visit the depot and wait for their items to be brought in by robots. The customer may pay when they arrive at the depot or may choose to wait for all items to arrive before doing so. After both are done, the transaction is complete, and the customer takes their items and goes on their merry way."*

## Application services

Starting with the four services—**Orders**, **Stores**, **Payments**, and **Depot**—on the right of *Figure 1.8*, we have the application services. These will all use events to communicate new states for triggers and notifications and will both publish them and subscribe to them. They will also have **GRPC application programming interfaces** (**APIs**) to support the API gateway layer.

## API gateway services

The API gateway layer displayed down the center of *Figure 1.8* will support a RESTful API for the customer kiosks, a management **user interface** (**UI**) with *WebSocket* subscriptions for the staff to use, and finally, a **gRPC** streams API for the robots. The API gateways are implemented as a demonstration of the **Backend for Frontend** (**BFF**) pattern.

The administrative BFF and the automation API gateways will create subscriptions to application events to allow delivery of state changes to clients. Note that we will not be developing API gateway services in this book.

## Clients

Finally, on the left of *Figure 1.8* are the expected clients, as outlined in more detail here:

- Customer kiosks, placed near or at mall entrances for ease of use
- An administrative web application for staff to manage the application data, process customer pickups, and take payment
- Shopper bot clients that perform autonomous shopping tasks for the busy customers

## A quick note about hexagons

You're going to be seeing a lot of hexagons in the diagrams of this book. The services in *Figure 1.8* all have some combinations of synchronous and asynchronous communication or connections, and all are drawn as hexagons, as depicted in the following diagram:

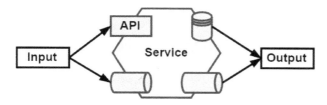

Figure 1.9 – Hexagonal representation of a service

The API gateway and application services are all represented as hexagons with inputs (such as the API and event subscriptions, shown on the left) and the outputs (the database and event publications, on the right). This is a visual presentation of hexagonal architecture, and we will be talking more about that in *Chapter 2, Supporting Patterns in Brief*.

# Benefits of EDA

An EDA brings several benefits to your application when compared to an application that uses only synchronous or **point-to-point** (**P2P**) communication patterns.

## Resiliency

In a P2P connection as shown in the following diagram, the calling component, **Orders**, is dependent on the called component, **Depot**, being available. If the called component cannot process the operation in time, or if the called component has a fault, then that error will propagate back to the caller. Worse is a chain or tree of calls that end up with a fault somewhere far away from the original caller, causing the entire operation to fail.

If the **Depot** service is not responding or is failing to respond on time, then the **Orders** service may fail to pass on information regarding new orders:

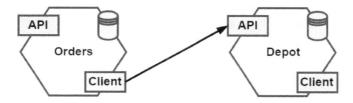

Figure 1.10 – P2P communication

In an EDA application, the components have been loosely coupled and will be set up with an event broker between them, as shown in the following diagram. A crash in an event consumer will have no impact on the event producer. Likewise, other faults (internal to the consumer) that cause it to temporarily be unable to process events again have no impact:

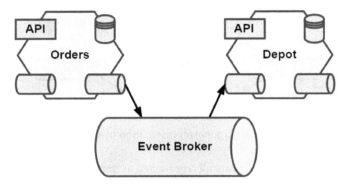

Figure 1.11 – Brokered event communication

Considering the example case of the **Depot** service becoming overrun with work, causing it to get backed up, orders submitted by the **Orders** service will be processed, just a little slower. The **Orders** service will be unaffected and continue to take orders as they come in. Better still, if the **Depot** service is down entirely, then it may only cause a longer delay until it can be restarted or replaced, and the **Orders** service continues.

## Agility

An event-driven application can be more agile in its development. Less coordination between teams is required when introducing new components to an application. The new feature team may drop in the new component without having to socialize any new API with any of the other teams.

The organization can more easily experiment with new features as an aside. A small team can stand up a new component without disrupting the work of other teams or the flow of existing processes. When the experiment is over, the team can just as easily remove the component from the application.

We can imagine that, at some point, an **Analytics** service could be introduced to the application. There are two ways this new service could be added. The first way is with a synchronous API (as shown in *Figure 1.12*) and the second is with an asynchronous event consumer (as shown in *Figure 1.13*).

When they choose to use the API, the team will need to coordinate with existing teams to potentially add new logic to capture data and new calls to their service. Completing this task will now require scheduling with one or more teams and will become dependent on them, as illustrated in the following diagram:

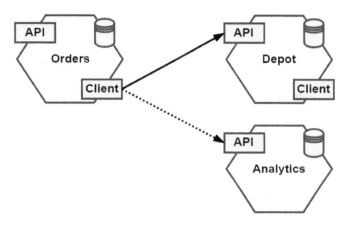

Figure 1.12 – New P2P service

Components that communicate using events make it easier for new components and processes to come online without requiring coordination with the teams in charge of existing components, as shown in the following diagram:

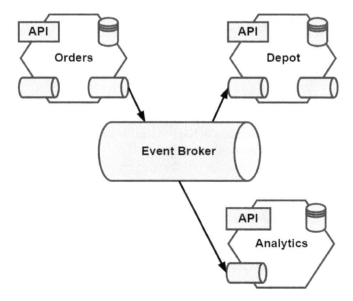

Figure 1.13 – New brokered event service

Now, when the **Analytics** service team has finished its work of picking which events to consume and captures the data that it needs, it can then add it to the application immediately.

If event streams are part of your EDA application, this also has the advantage of providing new components with a complete history of events to spin up with.

### User experience (UX)

With **Internet of Things (IoT)** devices exploding in number and millions of people having phones in their hands, users expect to be notified of the latest news and events the instant they happen. An event-driven application is already sending updates for orders, shipment notifications, and more. The organization may extend this to users more easily than a traditional synchronous-first application might allow.

### Analytics and auditing

Whether you're using event notifications, event-carried state transfer, or event sourcing, you will have ample opportunity to plug in auditing for the small changes that occur in your system. Likewise, if you're interested in building on analytics to your application to gather **business intelligence (BI)** for your marketing and product teams, often one or both are an afterthought, and in a traditional or non-EDA application, you may not have the data or can only recreate a partial picture.

## Challenges of EDA

Adopting EDA patterns for your application brings along some challenges that must be overcome for the application to succeed.

### Eventual consistency

Eventual consistency is a challenge for any distributed application. Changes in the application state may not be immediately available. Queries may produce stale results until the change has been fully recorded. An asynchronous application might have to deal with eventual consistency issues, but without a doubt, an event-driven application certainly will.

### Dual writes

Not entirely a challenge of event-driven applications alone, dual write refers to any time you're changing the application state in two or more places during an operation. For an event-driven application, this means we are making a change locally to a database, and then we're publishing an event either about the event or the event itself. If the events we intend to publish never make it to the event broker, then our state changes cannot be shared, and post-operation operations will never happen.

For this challenge, we have a solution that will have us publish our events into the database alongside the rest of the changes to keep the state change atomic.

This allows a second record of to-be-published events to be created, and even adds additional resiliency on top of what we got from using an event broker between components, as illustrated in the following diagram:

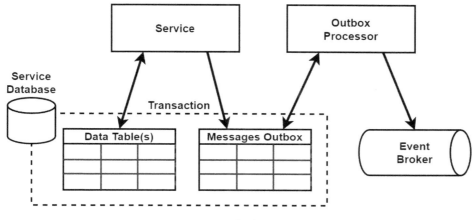

Figure 1.14 – Outbox pattern

We will learn more about this challenge and solution when I introduce you to the **Outbox** pattern in *Chapter 6, Asynchronous Connections*.

## Distributed and asynchronous workflows

Our third challenge involves performing complex workflows across components using events, making the workflow entirely asynchronous. When each component is coupled this way, we experience eventual consistency. Each component may not have the final state of the application when queried, but it will eventually.

This creates an issue for the UX and one for the collaboration of the components of the application involved with the operation. Each will need to be evaluated on its own to determine the correct solution for the problem.

### UX

The asynchronous nature of the operation would obviously make it difficult to return a final result to the user, so the choice becomes how to handle this limitation. Solutions include but are not limited to fetching the result using polling on the client, delivering the result asynchronously using WebSockets, or creating the expectation the user should check later for the result.

## Component collaboration

There are two patterns we can use to bring components together to manage workflows, as illustrated in the following diagram:

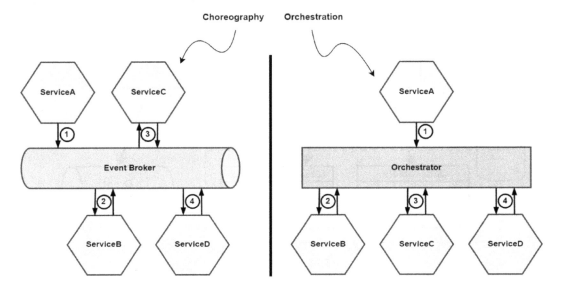

Figure 1.15 – Workflow choreography and orchestration

- **Choreography**: The components each individually know about the work they must do, and which step comes next
- **Orchestration**: The components know very little about their role and are called on to do their part by a centralized orchestrator

We will dive into the differences, some of the details to consider in choosing one over the other, and more in *Chapter 8, Message Workflows*.

## Debuggability

Synchronous communication or P2P involves a **caller** and **callee**. This method of communication has the advantage of always knowing what was called and what made the call. We can include a request ID or some other **unique ID (UID)** that is passed on to each callee.

One of the disadvantages of EDA is being able to publish an event and not necessarily knowing if anything is consuming that event and if anything is done with it. This creates a challenge in tracing an operation across the application components.

We might see multiple operations unrelated to one another spring up from the same event. The process to trace back to the originating event or request becomes harder as a result. For an event-driven application, the solution is to expand on the solution used for P2P-only applications, and we will see crumbs of this solution throughout the book and discuss it in more detail in *Chapter 12, Monitoring and Observability*.

Testing the application using several forms of tests will be covered in *Chapter 10, Testing*.

## Getting it right

It can be challenging for teams to think in terms of events and asynchronous interactions. Teams will need to look much more closely and know the application that they're building better to see the small details that sometimes make up events. In *Chapter 2, Supporting Patterns in Brief*, we will look at some patterns that teams can use to break down the complexities of an application, and how to make managing and maintaining event-driven applications easier in the long run.

In *Chapter 3, Design and Planning*, we will cover tools that teams can use to break down an application into behaviors and the events associated with each one.

### Big Ball of Mud with events

A **Big Ball of Mud (BBoM)** is an anti-pattern, where an application is haphazardly designed or planned. We can end up with one in our event-driven application just as easily with events as without and perhaps even more easily if we do not do a good job identifying behaviors and events.

## Summary

In this chapter, we were introduced to EDA, the types of events, and the core components involved with each event pattern you would find in an event-driven application. I covered some of the advantages from which you could benefit with an event-driven approach, and I introduced the challenges that will be encountered with this pattern.

In the next chapter, we will cover a range of patterns that will be used in the development of the demonstration application and why we might find them useful in conjunction with the development of an event-driven application.

# 2
# Supporting Patterns in Brief

There are a lot of software patterns we might use or come across in the development of an event-driven application. Event-driven architecture should not be the first tool you reach for in your toolbox.

We've been introduced to event-driven architectures, and now we'll see the patterns that work together with EDA to support excellent event-driven application design and development. These helpful patterns may not always be successful but using them in the right places and in moderation will improve your production time and reduce your bug rates.

In this chapter, we're going to cover the following main topics:

- Domain-driven design

- Domain-centric architectures

- Command and Query Responsibility Segregation

- Application architectures

## Domain-driven design

**Domain-driven design** (DDD) is a very large and complex topic, with entire books devoted to the use and implementation of the many patterns and methodologies that are brought together. I won't try to fit all of it into this chapter, much less this section, so we'll be taking a high-level look at the key strategic patterns that are useful to us as we design and develop event-driven applications. As for the tactical patterns, we'll be seeing examples of their use throughout the rest of the book.

> **Going deeper into DDD**
>
> For an in-depth look at DDD, I can recommend both *Domain-Driven Design: Tackling Complexity in the Heart of Software* by Eric Evans, as an original introduction to the topic, and *Implementing Domain-Driven Design* by Vaughn Vernon, for the expansion of the topic and a deeper dive into the strategic patterns of DDD. Finally, *Patterns, Principles, and Practices of Domain-Driven Design* by Scott Millett with Nick Tune rounds out the recommendations with its very deep and lengthy look at DDD.

## DDD misconceptions

The philosophies, methodologies, and patterns of DDD are well-suited for the development of event-driven applications. Before getting into DDD, I would like to cover a couple of misconceptions about it that developers might hold.

### Misconception one – DDD is a set of coding patterns

For most developers, their first exposure to DDD might be seeing an entity, value object, or some other pattern such as the repository that is being used in a code base they've worked on, or from some web tutorial covering a pattern or two. Regardless of the number of patterns they see, it is still an incomplete picture of what DDD is. Most DDD is never explicitly shown in the code, and a good amount of DDD comes into the picture before the first line is ever written.

### Misconception two – DDD is enterprise-level or leads to overengineered applications

DDD prescribes no specific architecture to use, and it neither instructs you how to organize your code for any given programming language nor enforces any rule that you must use in every corner of your application. DDD does not force you or your team to utilize a specific architecture, pattern, or code structure; that is something you are doing. The strategic patterns will actually assist you in identifying areas of the problem domain where you should not need to devote a lot of development time and resources.

Both misconceptions are centered around the use and a perceived overuse of the tactical patterns of DDD. As developers, we're technically minded people; we will search for a technical solution or a better way to do something when faced with a challenging or novel problem. What we've learned or used will find its way into our conversations when we include the names of the patterns. If all we seek out or share with others are the tactical patterns of DDD, then it's inevitable that we will miss out on the design philosophies and strategic patterns, only to turn around to complain that DDD has doomed another project.

# So, what is it all about then?

DDD is about modeling a complex business idea into software by developing a deep understanding of the problem domain. This understanding is then used to break up the problem into smaller, more manageable pieces. The two key patterns of DDD at play here are the **ubiquitous language** and **bounded contexts**.

## Alignment and agreement on the goals

To find success with DDD, collaboration must exist between domain experts and developers. There should be meetings where business ideas and concepts are sketched and diagrammed to be gone over from top to bottom and thoroughly discussed. The results of these discussions are then modeled and discussed further to weed out any incorrect understanding of implicit details.

This is not a process you do once before writing any code. Complex systems are living entities in a way, and they change and evolve. When new features are being considered, the same people should meet to discuss how these will be added to the domain model.

## Speaking the same language

When domain experts come together with developers, discussions could fall apart if the parties involved cannot come to an understanding of a concept by having different ideas about what is being said or read. The **Ubiquitous Language** (**UL**) principle requires every domain-specific term to have a single meaning within a bounded context. By using a shared language, a better understanding of the domain can flourish. The domain experts have their jargon and the developers theirs. It is preferable to use the terms spoken by the domain experts, and it is these terms that will be used to name and describe the domain models.

This is a core principle of DDD and a very important one too, but it doesn't come easy. Words that should be simple and have an obvious meaning may suddenly appear to have lost all meaning during discussions. Words may begin to develop a depth, which should highlight to everyone involved the importance of developing a UL and using it everywhere and always.

To hammer the point home, use the UL everywhere in code. It should drive the names of your function names, the structs, the variables, and the processes that you develop. When you sign off on the completion of some task or are given a bug to fix, the UL should always be used. This keeps the UL aligned across an organization.

When the UL is being spoken but confusion starts to appear, it could be a sign that the domain model is undergoing an evolution, and it might be a good time to have a meeting with the domain experts and developers again.

## Tackling the complexity

The complexity of the problem domain can be reduced by breaking the domain into subdomains so that we're dealing with more manageable chunks of the problem. Each new domain we identify falls into one of three types:

- **Core domains**: Critical components of the application that are unique or provide a competitive advantage to the business. These get the most focus, the most money, and the best developers. A core domain is not always obvious and can evolve or change with the business.

- **Supporting domains**: The utility components that provide functionality that supports the core business. You might consider using an off-the-shelf solution if what is being provided and developed by a team is not specific enough to the business.

- **Generic domains**: Components that are unrelated to the core business but necessary for it to function. Email, payment processing, reporting, and other common commodity solutions fall into this domain type. It wouldn't make sense to devote teams to develop this functionality when so many solutions exist.

As a business changes in response to competition or other factors, it is possible over time for the type associated with a domain to change or for the domain to split into two or more new domains.

Using a **core domain chart** to chart the business differentiation and model complexity for each domain in our *MallBots* application, we end up with the following:

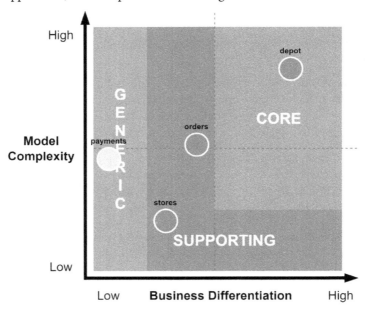

Figure 2.1 – A core domain chart for the MallBots domains

In *Figure 2.1*, we've identified that the **depot** has the highest value to the business, is going to be rather complex, and is going to be our **core** domain. Taking **orders** and managing the **store's** inventory is important to the business, but it has no differentiators and provides **supporting** functionality only. Payments exist simply because they must, so we've decided to integrate with a third-party SaaS to handle our money, which makes our last domain **generic**.

## Modeling

The **domain model** is a product of the collaboration between domain experts and developers using the UL. What goes into the model should be limited to the data and behaviors that are relevant to the problem domain, not everything possible in an attempt at modeling reality. The point of a domain model is to solve problems identified in the domain.

Eric Evans suggests experimenting with several models and not getting stuck too long on minutia. You are trying to pull out what is important from the conversation with the domain experts. Listen for connecting words to identity processes and behaviors, titles and positions to identify actors, and, of course, the names of things to identify data. This should be captured on a large surface such as a whiteboard or a large roll of paper or a blank wall if you're doing **EventStorming**. We will talk more about using EventStorming as a method to develop a domain model more in *Chapter 3*, *Design and Planning*.

The model should be free of any technical complexities or concerns, such as mentioning any databases or inter-process communication methods and should only be focused on the problem domain.

## Defining boundaries

Every model belongs to a **bounded context**, which is a component of the application. Because the model belongs to this context, care needs to be taken in keeping it safe from outside influences or enabling external control.

You have broken down the complexity into multiple domains and discovered the models hidden within your software. The boundaries that we see forming from our discovery efforts will be around the business capabilities in our application. Examples of business capabilities for the *MallBots* application are the following:

- Order management
- Payment processing
- Depot operations
- Store inventory management

All the domains should not have a singular view of any given model; they should be concerned with the parts that are relevant to a particular bounded context.

Every bounded context has its own UL, which should be taken to mean terms that might have different meanings when contexts change across an application. The products that are picked out by a customer will exist in several domains and, depending on the context, have completely different models, with different purposes and attributes. When the domain experts and developers discuss *products*, they will need to include the context to which they're referring. They could be talking about the inventory for a store, the line items in an order, or fulfillment and delivery at the depot.

A bounded context takes on a technical aspect in that its implementations introduce some technical boundaries around the models. For a distributed application, a bounded context typically takes on the implementation of a module or a microservice, but not always. A very distinct boundary exists where the context limits the mutations and queries of the model it has been created to maintain.

> **Bounded contexts are a difficult concept**
>
> To do DDD well, you must understand bounded contexts. I encourage you to read one of the suggested books or do a search and learn more about them. Finding or determining the right boundaries in an application is not a science and is very much an art.

## *Tying it back together*

It may seem counterintuitive that so much effort is expended breaking down our problem domain into smaller domains and bounded contexts, only to later design how they should all interact again. The bounded contexts and their high walls now need to be made to work together and become integrated again. We use **context mapping** to draw the relationships between our models and contexts that we'll need for our application to be functional.

The purpose of context mapping is to recognize the relationships the models will have with other models and to also show the relationship between teams. The patterns used in context mapping are of a descriptive value only. They do not give any hints about what technical implementations should exist to connect the models:

- Upstream patterns:

  - **Open host service**: This context provides an exposed contract that downstream contexts may connect to

  - **Event publisher**: This context publishes integration events that downstream contexts may subscribe to

- Midway patterns:

  - **Shared kernel**: Two teams share a subset of the domain model and maybe the database.

  - **Published language**: A good document shared language to translate models between contexts. It is often combined with an open host service.

- **Separate ways**: Contexts that have no connections because integration is too expensive.

- **Partnership**: A cooperative relationship between two contexts with joint management of the integration.

• Downstream patterns:

- **Customer/supplier**: A relationship where the downstream context may veto or negotiate changes to the upstream context

- **Conformist**: The downstream service is coupled with the upstream context's model

- **Anticorruption layer**: A layer to isolate the downstream context from changes in the upstream context's model

Applying the preceding patterns to our application, we could end up with the following:

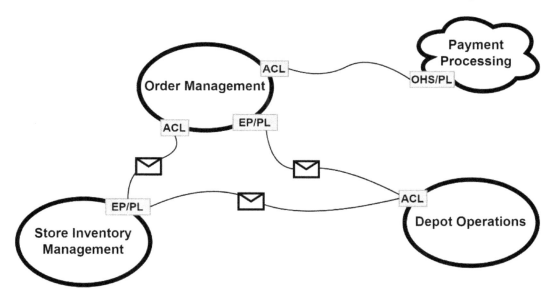

Figure 2.2 – A context mapping example

## How is it useful for EDA?

DDD is generally useful for event-driven applications, and you may do just fine without it. What it brings to the table, in terms of digging into the business problem with the domain experts and developing a UL to break down the complexity into bounded contexts, cannot be overlooked.

Event-driven applications will benefit from making the efforts to create better event names, by determining which events are integration events and will become part of the contract for a bounded context.

# Domain-centric architectures

A **domain-centric architecture**, to reiterate, is an architecture with the domain at the center. Around the domain is a layer for application logic, and then around that is a layer for the infrastructure or external concerns. The purpose of the architecture is to keep the domain free of any outside influences such as database specifics or framework concerns.

Before we discuss more about domain-centric architectures, let's first look at some traditional, or enterprise, architectures.

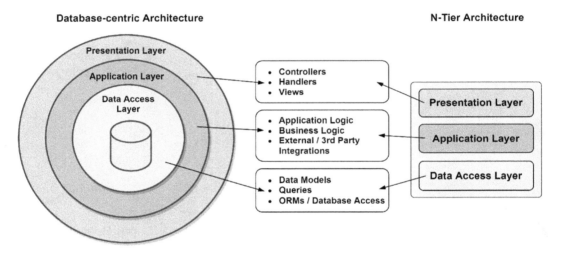

Figure 2.3 – Some traditional architectures

The problem teams will notice with traditional architectures is that, over time, the cost to maintain the application increases. These architectures are also hard to update when infrastructure choices or requirements have changed. In both architectures from *Figure 2.3*, the applications are broken into layers and are not much different conceptually. It isn't the layers that are the cause of the issues; it is how they are tightly coupled together. The data models from the **data access layer** are used in the **application layer** and the **presentation layer**. The reverse can also be true; the UI frameworks will have their request models used in the other layers.

As a result, each of the three layers becomes dependent on the models used in the other two layers. Having these dependencies will mean that a change in the **presentation layer** is likely going to require a change in the **data access layer**. Dealing with the tangled web of dependencies and the tight coupling will result in an organization's resource expenditures increasing more rapidly over time.

## An evolving solution

**Alistair Cockburn** invented **hexagonal architecture** in 2005 while explaining the **ports and adapters** pattern applied to application design (`https://alistair.cockburn.us/hexagonal-architecture`), as a solution to the tight coupling made between the parts of an application:

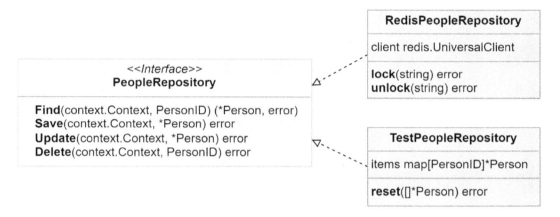

Figure 2.4 – A port and two adapters

In his solution, there were two parts – the inside of the application and the outside. Each outside dependency the application used would be broken up into two parts – a **port** or interface, and an **adapter** or implementation. The `PeopleRepository` interface in *Figure 2.4* represents a port, and `RedisPeopleRepository` and `TestPeopleRepository` both represent adapters that implement the interface. Using this technique, our applications will now be isolated from the changes made to the outside dependencies.

Then, in 2008, **Jeffrey Polermo** introduced us to his **onion architecture** (`https://jeffreypalermo.com/2008/07/the-onion-architecture-part-1/`). The **dependency inversion principle** (more on that in a later section) would play a large role in this new architecture:

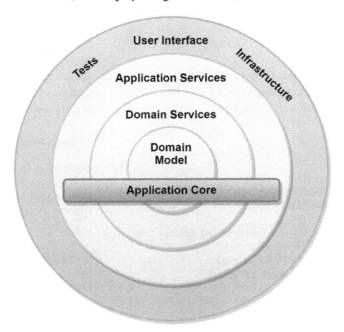

Figure 2.5 – The onion architecture

An application using hexagonal architecture is now broken up into different layers, the **application services**, the **domain services**, and the **domain model**. The external dependencies create the outermost layer around the **application core**. Dependencies point inward toward the **domain model**, and the outer circles contain implementations of the interfaces located in the inner circles.

Palermo also suggested the use of an **inversion of control** container to handle the work of dependency injection. Go does not have great language support for dependency injection, but we will see some possible solutions in *Chapter 4, Event Foundations*.

**Robert C. Martin** made a post in 2012 after studying hexagonal architecture and onion architecture, along with some others, to introduce **clean architecture** (`https://blog.cleancoder.com/uncle-bob/2012/08/13/the-clean-architecture.html`).

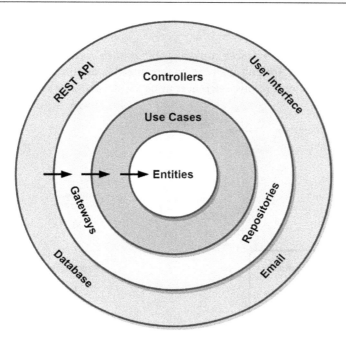

Figure 2.6 – The clean architecture

He noted that the architectures had many similarities:

- None relied on any frameworks to scaffold applications on top of
- The designs produced more testable code
- Infrastructure was viewed as a substitutable dependency

## Martin's dependency rule

He declared that *"source code dependencies can only point inwards"* as the most important aspect of domain-centric architectures. Nothing at all can be referenced in an inner circle that existed in an outer circle. The application must resolve references using the dependency inversion principle.

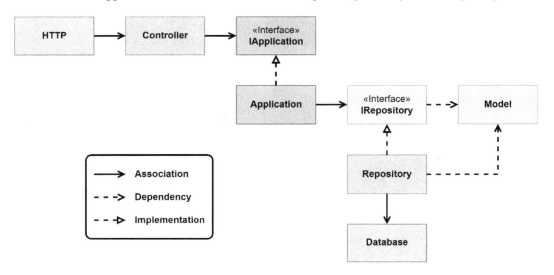

Figure 2.7 – The Dependency Inversion Principle

Starting with the most isolated items and referencing the layers from *Figure 2.6*, we have the **model** and **IRepository** interface. These may not have any references to anything else, only to other entities in the **Entities** layer. Next is the **application** and the **IApplication** interface, which belong to the **Use Cases** layer and may only use items in the Use Cases or Entities layers. From the Interface Adapters layer, we have the **controller** and **repository** implementations. Our last layer might appear to contain an error; the association arrow is pointing to the **database** from the **repository** when, according to the **Dependency Inversion Principle** (**DIP**), it should not. While the concrete implementation will contain a reference to the database to work, the **IRepository** interface will keep the **application** isolated from any specific database implementation.

# Hexagonal architecture applied

**Alistair Cockburn** invented hexagonal architecture to address the spread of business logic into other unrelated parts of the software. He laid out three factors of this problem:

- Testing is more difficult when the tests become dependent on the user interface

- Coupling makes it impossible to shift between human-driven use and a machine-driven one

- Switching to new infrastructure is difficult or impossible when the need or opportunity presents itself

The solution was to isolate the application and its core from external concerns by placing APIs, the **ports**, on the boundary of the application that used **adapters** to integrate with external components. This pairing of abstraction and concrete implementations would allow external components such as new UIs, test harnesses with mocks, and new infrastructure to be swapped in and out much more easily.

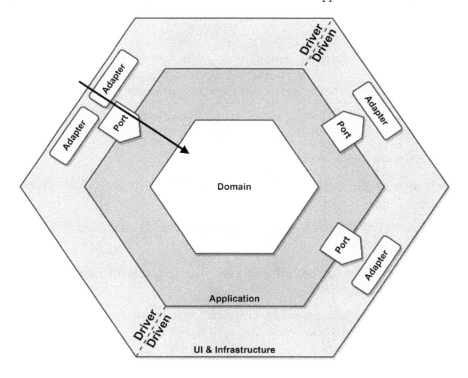

Figure 2.8 – An interpretation of hexagonal architecture with elements of clean architecture

The early diagrams of hexagonal architecture didn't pay as much attention to the domain that clean architecture did. In *Figure 2.8*, I've added a **domain** to the application and a **UI and infrastructure** hexagon to create a blended interpretation of the two architectures.

### Domain

In the center of the diagram, we have our **domain**. This layer of our application contains our domain model, domain-specific logic, and services. This layer is the least affected when external changes are made.

This layer of the application has no other dependencies and is free of any references to external concerns or application services.

### Application

Surrounding the **domain** is the **application** layer that contains our application-specific logic and services. The application layer will also define the interfaces that external concerns will be using to interact with the application.

The application layer may only ever depend on the **domain** layer and cannot reference external concerns.

### Ports and adapters

Outside of the application are all external concerns. We'll find the frameworks, UI implementations, and databases for saving our data. Everything outside of the application interacts with it using a port. The port is an abstraction known to the application that allows it to use and be used by external concerns.

In the other half of the interaction, the adapter is some small piece of code that intimately knows how to communicate with the external dependency.

These pairs of ports and adapters come in two types:

- **Driver** or primary adapters are the web UIs, APIs, and event consumers that drive information in our application
- **Driven** or secondary adapters are the database, loggers, and event producers that are driven by the application with some information

While they are typically paired up, that isn't always the case, and you might have a situation where more than one adapter is using a port.

Communication between the adapters and the application happens only through the ports and the **Data Transfer Objects (DTOs)** that they have created to represent the requests and responses.

## Testing

The abstractions we've used to isolate our application and domain model from external concerns will also help us in testing. A test harness can take the place of any primary adapter to execute tests of the application. We can also use a mock application to test real database calls for integration testing.

The architecture and the separation of concerns forced on us from the layers have resulted in us writing smaller components. By extension, we've written more testable components as a result.

## A rulebook, not a guidebook

Domain-centric architectures provide the rules for writing better code, not a guide for how to do that exactly. I'm talking about how you organize your packages and modules in Go, how you will write your constructor functions, or what method you use for dependency injection.

## Should you use domain-centric architectures?

Is testing important to you? What about maintainability? A domain-centric architecture application will be highly testable and be cheaper to maintain in the long run. A sufficiently large application, and especially one that is using DDD, will see more benefits from using a domain-centric architecture than drawbacks.

Having your application core independent of framework or infrastructure choices, and any vendor lock-in such as cloud provider dependencies, also gives it a high degree of portability and reuse.

### What about those drawbacks?

A domain-centric architecture will require a larger investment upfront and is going to be a challenge for the less experienced developers. In the eyes of some engineers, the requirements or constraints of domain-centric architectures can cause an application to be bloated or over-engineered. To some developers, needing to maintain abstractions for every dependency or using dependency injection adds needless boilerplate code and more work.

Like DDD, an implementation of domain-centric architectures can go south if they are followed rigidly, and worse if the interpretation is wrong and the wrong choices are being made. Developers will become discouraged, and the project may be counted as another victim of overcomplication blamed on the architecture.

## How is it useful for EDA?

Domain-centric architectures are also generally useful, and you might skip using them if you keep your services small enough or never have to deal with migrating cloud providers or switching databases.

# Command and Query Responsibility Segregation

**Command and Query Responsibility Segregation** (**CQRS**) is a simple pattern to define. Objects are split into two new objects, with one being responsible for commands and the other responsible for queries.

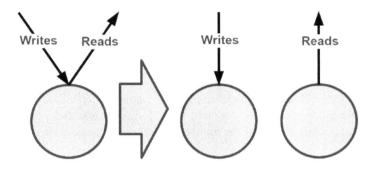

Figure 2.9 – Applying CQRS to an object

*Figure 2.9* demonstrates just how simple the concept might be, but the devil is in the *implementation* details as they say. The definitions for Command and Query are the same as they are for **Command-Query Separation** (**CQS**):

- **Command**: Performs a mutation of the application state
- **Query**: Returns application state to the caller

> **Note**
> In CQRS, just as it is in CQS, an action can either be a **command** or a **query** but not both.

## The problem being solved

The domain models we've developed with the help of domain experts may be very complex and large. These complex models may not be useful or too much for our queries. Conversely, we may have complex queries that make us consider modifying our domain models to support, which may be a violation of our UL. We also may be unable to serve a query with the domain model we have ended up with.

## Applying CQRS

An analogy I use to describe applying CQRS is to visualize your application like a ribbon:

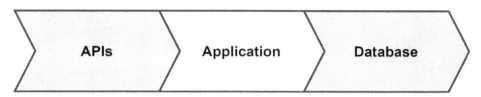

Figure 2.10 – A simple application ribbon

This application, shown as a ribbon in *Figure 2.10*, can be cut horizontally, creating a top side and a bottom side at any point. Where you make the cut and how far will determine how much of the CQRS pattern you're applying to your application.

### Applied to the application

You might want to apply CQRS to the application code only:

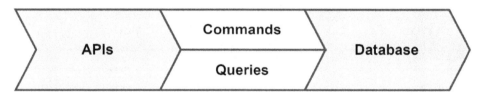

Figure 2.11 – CQRS applied to an application

With an **application** divided into a **command** side and a **query** side, you can apply different security models to each side or decide to reduce the complexity of your service objects. You may continue to use the same **database** but use an ORM on one side and raw SQL for performance purposes on the other. This would arguably be the least effective use of CQRS that you can apply to your application.

### Applied to the database

You can extend your use of CQRS to the database:

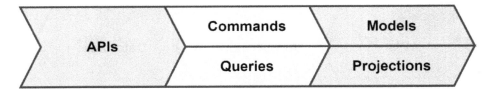

Figure 2.12 – CQRS applied to the database

Fined-tuned SQL queries will only get you so far. Moving your queries over to a new data store such as a NoSQL, key-value, document, or graph database may be necessary to keep up with the load. You can utilize an event-driven approach to populate multiple new projections within multiple services.

### Applied to the service

Cut all the way through, you split the service into two:

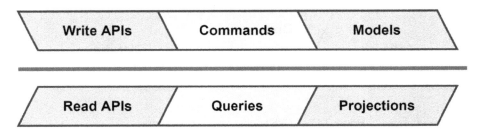

Figure 2.13 – CQRS applied to the service

Applying CQRS to the entire service gets you two services that can be scaled separately; they can be maintained by different teams and have entirely different technology stacks.

## When to consider CQRS

Let's explore the points while considering CQRS:

- Your system experiences a much larger amount of read operations than write operations. Using CQRS allows you to break the operations into different services, allowing them to be scaled independently.
- Security is applied differently for writes and reads; it also limits what data is viewable.

- You are using event-driven patterns such as event sourcing. By publishing the events used for your event-sourced models, you can generate as many projections as necessary to handle your queries.

- You have complex read patterns that bloat or complicate the model. Moving read models out of the domain model allows you to optimize the read models or the storage used for each access pattern.

- You want the data to be available when writing is not possible. Whether by choice or not, having the reads work when the writes are disabled allows the state of the application to still be returned.

## CQRS and event sourcing

**CQRS** is not, in my opinion, an event-driven pattern. It can be used entirely without any kind of events or asynchronous approaches. It is, however, very common to hear it talked about alongside **Event Sourcing**, and that is because the two work well together.

One of the benefits of splitting your model into two parts is that your write side is reduced to writing to an append-only log, and another benefit is that you are free to have as many read models as you need that are fed by the same events. These read models can be built for very specific needs and spread out across your application.

## Task-based UI

One of the goals of CQRS is to make the behaviors that drive the commands that your application executes explicit on the write side. That is difficult to do when an application is driven by a **Create, Read, Update, and Delete** (**CRUD**) UI. The intended behavior of a user's action is frequently lost behind the usage of basic commands such as `UpdateUser` in this type of UI. Supposing that call was also used when the user updated their profile, when they changed their mailing address, it would be difficult to determine which was the intended action.

By using a task-based UI, where each action has a clear intention, we can communicate the user's intended behavior more clearly. Now, when the profile is being updated, the UI would make a call to the `UpdateProfile` API, and when the mailing address changes when the customer has moved, it would call the API with `ChangeMailingAddress`.

# Application architectures

For an event-driven application, there are a few application architectures that we can choose between. They have their pros and cons, and for green field projects, there is only one recommendation I'd make.

## Monolithic architecture

This is an application that is typically built from a single code base and is deployed as a single resource. These kinds of applications have the advantage of being easy to deploy and are relatively simple to manage and operate. Outside of needing to maybe communicate with some third-party APIs, a single user interface and database will be most of the infrastructure concerns. The application shown in *Figure 2.14* is easy to scale to handle more users by simply deploying it to more instances that point to the same database:

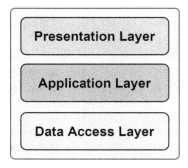

Figure 2.14 – A monolith application

On the other hand, the larger a monolith grows, the harder it is for teams to develop it efficiently, as the development of new features sees them come into conflict and constant deployment becomes a faint memory. The architecture also gets an unfair amount of negativity regarding the messy code that goes into the development of a monolith. That negativity is unfair because that can happen with any code base and has to do with bad design.

### *Modular monolith architecture*

The modular monolith shares a lot of the benefits and drawbacks of a monolithic architecture but also shares a good number of the advantages of a microservices architecture, with only a few of the drawbacks.

If we apply DDD and a domain-centric architecture to our existing monolithic application, we can refactor it toward a modular monolith architecture. By identifying the domains of our application and defining bounded contexts, we can split the core of the monolith into however many modules we need to.

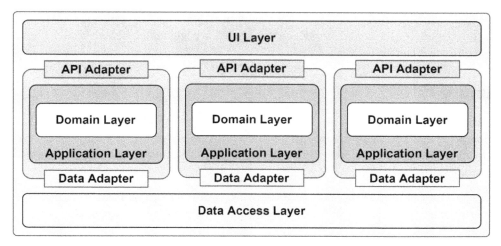

Figure 2.15 – Modular monolith

Our refactored application shown in *Figure 2.15* is now built with three modules that can be more independently worked on by different developers or teams.

Any communication between the modules should be treated like any other external concern and used as an interface and concrete implementation to support an enforceable contract.

## Microservices

A microservices architecture involves building individual services that are ideally aligned with a bounded context to create a distributed application. The advantages of microservices over a monolithic application are that they're independently deployable and can be independently scaled. They also have better application resiliency thanks to fault isolation. The advantages of being loosely coupled might be an advantage over a poorly designed monolith but not over a modular monolith. Individually, the services will be smaller code bases and easier to test.

Microservices have several drawbacks as well. Foremost is the complexity involved with managing many cooperating but independent services. Eventual consistency, which is largely caused by the architecture's distributed nature, must also always be taken into consideration. Performing larger tests may involve multiple microservices, making the effort more complicated.

## Recommendation for green field projects

A modular monolith is the recommended architecture to start with for any project of reasonable complexity. A team will be able to better focus on the domain model implementation and not necessarily require additional external support to deploy an application.

After the application has outgrown the module monolith architecture, the team will be able to very easily extract the modules into microservices when needed to begin taking advantage of the benefits of the microservices architecture.

## Summary

In this chapter, we took a look at some of the key strategic patterns of DDD and how they're used to develop better applications. We were also introduced to domain-centric applications as ways we might organize our applications after working so hard to develop the right bounded contexts and domain models.

We then looked at CQRS and how its simple pattern can be used alongside event sourcing to create a more performant application. Finally, we covered application architectures that would benefit from the patterns of EDAs.

In the next chapter, we will discuss and use some tools to design and plan the *MallBots* application.

## Further reading

- *Domain-Driven Design Reference* by Eric Evans: `https://www.domainlanguage.com/wp-content/uploads/2016/05/DDD_Reference_2015-03.pdf`

- *CQRS Documents* by Greg Young: `https://cqrs.files.wordpress.com/2010/11/cqrs_documents.pdf`

- *Modular Monolith: A Primer* by Kamil Grzybek: `https://www.kamilgrzybek.com/design/modular-monolith-primer/`

# 3
# Design and Planning

It is now time to put into practice what we talked about in the previous two chapters. As the saying goes, before we can run, we must learn to walk, and before we can walk, we must learn to crawl. We were introduced to the *MallBots* application back in *Chapter 1, Introduction to Event-Driven Architectures*, but before we can create that application, we must have a plan built on a better understanding of the problem the application is intending to solve.

In this chapter, we will cover the following topics:

- What are we building?
- Finding answers with EventStorming
- Understanding the business
- Recording architectural decisions

We will be using **Domain-Driven Design (DDD)** discovery and strategic patterns as the basis for our initial approach. To facilitate the discovery, a workshop technique called **EventStorming** will be used to organize meetings with domain experts and developers. The knowledge we gain from these meetings about our application will also be used to design specifications that will be used to perform acceptance testing later and throughout the book.

Toward the end of the chapter, we will use the tactical patterns of DDD to design the models and behaviors in more concrete terms that will lead us toward a prototype.

# Technical requirements

You will need to install or have installed the following software to run the application or to try the examples:

- The Go programming language
- Docker
- The source code for the version of the application used in this chapter can be found at `https://github.com/PacktPublishing/Event-Driven-Architecture-in-Golang/tree/main/Chapter03`

# What are we building?

If you recall the *MallBots* application pitch from *Chapter 1, Introduction to Event-Driven Architectures*, we are building an application that is not a typical e-commerce web application but not too far removed from one either. Just before the pitch, a diagram was also shared that showed a very high-level view of what the final application would be comprised of. Getting from the pitch to a final application can happen in any number of ways. If you were to take those two bits of information and sit down to immediately started writing the code, where would you even start? Let's see.

We will use the following process to arrive at a design for our application:

1. Use EventStorming to discover the bounded contexts and related ubiquitous languages
2. Capture the capabilities of each bounded context as executable specifications
3. Make architectural design decisions on how we will implement the bounded contexts

# Finding answers with EventStorming

Getting knowledge from domain experts to developers could take several meetings. No one enjoys attending meetings that are either boring or non-conclusive. A sit-down meeting between developers, who will be asking a lot of questions and have some assumptions, and domain experts, who have the answers, could go down a rabbit hole on a single issue that has a small portion of the attendees involved.

Normal meeting etiquette is to avoid side conversations, which would waste the time of all the people not involved in those discussions. We do not want to use a meeting format that forces a group to focus on one issue after another; we should prefer a workshop format that encourages multiple conversations on issues and topics at once, such as EventStorming.

## What is EventStorming?

EventStorming is a fun and engaging workshop that uses colorful sticky notes to quickly visualize the building blocks of the flows that make up your application. It intends to uncover as many of the implicit details locked away in the heads of a few people and share that knowledge with domain experts and developers alike. The workshop is made up of a series of steps that expand on the work that came before to build a visual representation of a domain or problem, as shown in the following figure.

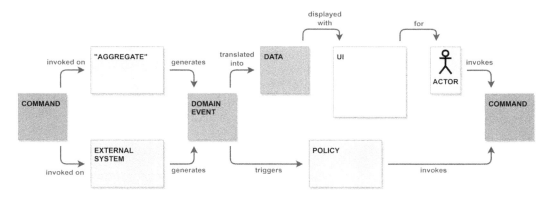

Figure 3.1 – A flow diagram using EventStorming concepts

Let's look at the EventStorming concepts shown in *Figure 3.1*:

- A **domain event (orange)** represents something that has an impact on your system and may occur either inside or outside. It should be written in the past tense.

- A **command (blue)** is an action or decision that is invoked by users or policies.

- A **policy (lilac)** is a business rule that can be identified by listening for the phrase *"whenever <x>, then <y>."*

- An **aggregate (tan)** is a group of domain entities acted on as a single unit.

- An **external system (pink)** is a third-party system that is external to an application. It may also represent other departments not involved with the workshop or internal services we do not control or maintain.

- **Data (green)** is information recorded in the system or is the required information for commands and business rules.

- An **actor (yellow)** is a user, role, or persona that creates actions in our system. It includes a drawing of a stick figure or another simple representation of the actor. For persona or role actors, you can include a distinctive hat to help differentiate one actor from another.

- A **UI (white)** is an interface example or screen mockup.

There are more concepts to EventStorming than those presented in the *Figure 3.1* flow diagram:

- **Definitions** for words and phrases that are part of the ubiquitous language used by the business.
- **Hotspots** for questions or to point out problems.
- **Opportunity** stickies can be placed to create a future call to action.
- **Happy** path and **unhappy** path stickies are used to label outcomes of branches in a flow. You are free to deviate from the suggested colors and sizes of sticky notes and to come up with additional concepts that help with *storming* out your particular problem.

You may recognize several names that EventStorming has in common with DDD and that is not an accident. Big Picture EventStorming focuses on the discovery of the bounded contexts and the ubiquitous languages by looking at the entirety of a business problem. Organizations can follow up a Big Picture workshop with design-level EventStorming workshops that dive into different complex or core contexts, in order to model them using tactical DDD patterns.

---

**Why use sticky notes?**

They can be very easily moved around while drawings on a whiteboard cannot, at least not easily. Additionally, stickies can be stuck on top of other notes to create associations that can then be moved as a group. It also allows more people to stand up and get involved with a workshop, because they are only adding a small note to any available space and do not have to fight for a region in which to draw or diagram their ideas.

---

## Big Picture EventStorming

We do not have a wealth of knowledge to draw on for this application and will need to develop it. The best EventStorming format for us, in this case, will be Big Picture. This format of EventStorming uses only the domain event stickies at the beginning and introduces just a few more as the workshop advances through its steps, in order to not overwhelm the participants with too much. The goals of Big Picture are discovery and knowledge transfer.

---

**A fictitious workshop example**

To explain EventStorming, we will run through a fictitious workshop attended by the developers and domain experts working on the *MallBots* application.

---

When we meet with the company for the first time to discuss their application, we ask them to bring along key people from across the company. From our side, we also bring key architects and developers to keep the total head count reasonably low. During the sessions with the company, we will be focused on what the application will do and not how it will do it. We are not interested in making any decisions or assumptions on technologies such as the web server, the cloud provider, the databases, or how things are going to be implemented.

Big Picture EventStorming is broken up into a series of steps:

1.  **A kick-off** is a quick session to introduce everyone to a workshop, its goals, and each other.

2.  **Chaotic exploration** is a discovery of all the events that happen in an application.

3.  **Enforcing the timeline** is about bringing order to the chaos and also identifying the source of our events.

4.  **People and systems** involve identifying the people causing events in our timeline and any external systems we interact with.

5.  An **explicit walk-through** has participants taking turns narrating portions of the timeline.

6.  **Problems and opportunities** is a call for everyone to share their opinions on issues and ideas.

## Step one – kick-off

During the kick-off, the facilitator should, if they don't already know, take a poll to determine how many participants may need or want an introduction to the EventStorming workshop process. Examples for performing a quick introduction include a time-limited EventStorming of popular movies or books. The facilitator will also put forth and explain the goals for the group during the workshop.

Now is as good a time as any to introduce some tips and etiquettes of EventStorming:

-   Do not move or replace any sticky notes without first discussing the action with the writer. There is only so much room on a little sticky note, so try to avoid making assumptions about the meaning behind a few words.

-   There is no harm in guessing. Put up a sticky note if you think it is important and relevant to the business problem. After it is up, you can then ask any questions to put to rest your doubts. Don't ask before putting something up on the timeline. Others will be able to give better ideas and feedback if they see the event up on the timeline.

-   Do not get attached to what you have written. Replace stickies with ones with better wording to avoid ambiguity, and to adopt the ubiquitous language used by domain experts.

- During chaotic exploration, everything stays. If you find yourself second-guessing something you have placed on the timeline, fight the urge to crumple up and remove the stickies. Your first thought might be going in the right direction, but you got some terminology incorrect, and it just needs to be reworded. You may also find it fits in better elsewhere on the timeline and could be moved. If you cannot decide what to do, just move the sticky note somewhere isolated, such as below the timeline, and it can be discussed during a break or in a later step.

- Take a couple of steps back to think. By stepping back, you will be able to get a wider view of the timeline and the opportunity to see what else is being put up. It will also give others a less obstructed view of the timeline when they are also doing their thinking.

### Step two – chaotic exploration

We'll start the process by focusing on the **domain events** that happen across an application. Everyone is given a pad of orange sticky notes and a marker, and somewhere in the room are spares of both. Participants will think of as many events as they can and then make a guess as to where they should be placed on the timeline. The events we're interested in are going to be relevant to domain experts, not on any implementation or technical details. Domain experts are interested in products being placed into a shopping basket, not records saved into a database.

It may be helpful for a facilitator to get the ball rolling by putting up one or a couple of events as examples. The facilitator is there to support the participants and not to lead them through the session. At most, the session for this step should take 1 to 2 hours.

Each participant will work independently to choose which events should be included on the timeline. Participants should avoid attempting to reach an agreement on the sticky notes that they are placing on the timeline. The purpose of this step is to identify the events that take place in chronological order. If a participant is stuck thinking of new events, the facilitator may suggest they pick an initial event and work on determining the events that come before or after it.

The facilitator should break up the sessions to keep the minds of the participants sharp and focused. We want to keep activity high and the momentum going, and when participants are slowing down in both respects after already taking a break, then as the facilitator, you should call an end to the session. The goal of exploration is to produce output from the discovery of significant events, not to consume everyone's time.

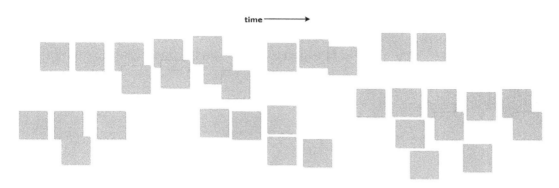

Figure 3.2 – Chaotic exploration results

**Bigger Picture EventStorming**

Fitting the entire timelines on the page would result in an illegible representation of sticky notes, so you'll find full-sized images of the timelines, with text on each sticky note representing each step, in the `ch3/docs/EventStorming/BigPicture` folder in the source code repository.

During this first step, participants are likely to encounter situations where an event that another participant has placed on the timeline does not make sense to them. Not everyone will come into the workshop knowing all of the terminology used by other departments, and the developers might not know the complete business language the domain experts are using. When we encounter confusion regarding words or phrases, a **definition** sticky note can be created and placed somewhere near the timeline. These definitions will help build our ubiquitous language.

Let's look at definitions in the following points:

- **Store**: A physical store in the same mall as the *MallBots* service. We track the name and location.

- **Participating store**: A store that has been approved and can be selected for automated shopping.

- **Catalog**: Store items that have been made available for purchase with the service. We track the name, price, and picture of the items.

- **Cart**: A customer's store and product selections that have not yet been submitted as an order.

- **Order**: A customer-submitted request for items to be automatically shopped for and collected at the depot.

The timeline of the first EventStorming session for the *MallBots* application is shown in *Figure 3.2*. The results are a mostly disorganized timeline of events with sloppy grouping. This is a reasonable result to see at the end of step one. Other outcomes can include numerous duplicate events, and we could have entire flows modeled more than once from different perspectives. Missing events at this point aren't going to be the end of the world either, especially given the time-boxed nature of the session.

It is messy, but we can start to see the different parts of the application take shape. The cart flow found at the top left of the timeline appears to be mostly complete, but the bot and depot have received less attention in the bottom-right corner. Store management down at the bottom left of the timeline is an example of how some flows might receive very little attention. This could mean some key people were missing from the workshop, or that part of the application is not considered to be that critical to the success of the business.

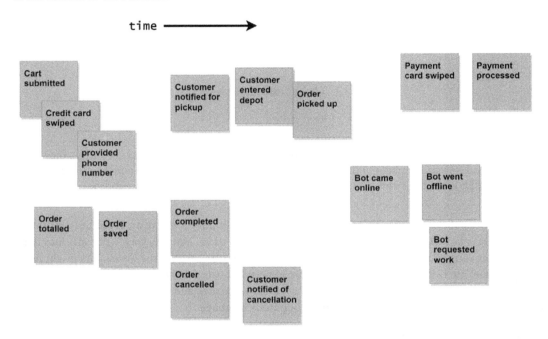

Figure 3.3 – A close-up of the chaotic exploration results

As we can see in *Figure 3.3*, the events associated with the orders are spread out into two groups, with most of them at the bottom left of the close-up view. There are also a couple of notifications added, but they are not grouped; instead, they are put with other events that they seem to be associated with. Remember that participants should try to place events in chronological order but shouldn't waste time trying to get groupings accurate.

When the session is over, and everyone has stepped back from the wall, it would be a good opportunity for participants to discuss their observations of the workshop and the timeline results before taking a break before the next step.

## Step three – enforcing the timeline

The next step in the workshop process is to organize the events into their correct chronological order and to group related events together. Grouped events are called flows, and each flow should represent a process belonging to a domain. We will be modeling multiple flows and want to keep the flow of events in any parallel flows in sync. Organizing the events into flows will start with the expected path, or happy path, for a process. After the happy path has been organized, we can begin to add branches for the alternative or unhappy paths that can result from bad user input or errors occurring in the application.

The facilitator will now take charge of the room and will be the one either moving the sticky notes around after some discussion or instructing the participants as groups to organize portions of the timeline. The purpose of this step is to bring order to the chaos we have allowed to happen in the previous step, and individual efforts might be counterproductive for that to occur.

There are multiple strategies we can use here to add structure to the timeline. Example organization methods include but are not limited to the following:

- **Pivotal events**: An organizational method that identifies significant events in a timeline to split the timeline vertically. These would be represented with larger sticky notes and a vertical divider, made with tape or a marker, that runs under the event.

- **Swim lanes**: The method of using horizontal dividers along the timeline to split events into flows that belong to specific actors in our application.

- **Temporal milestones**: Like the pivotal events method but uses time instead of events to split the timeline.

- **Chapter sorting**: Useful for organizing timelines with an overwhelming number of events and/ or a limited amount of space. Identify the chapters of events, organize those, and then go back and organize the events for each chapter.

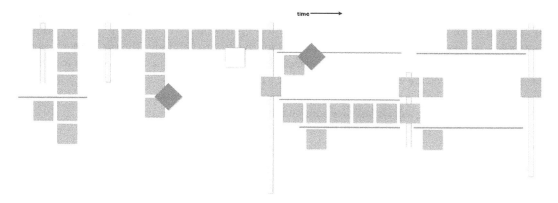

Figure 3.4 – Enforcing the timeline results

In *Figure 3.4*, we have used both pivotal events and swim lanes to organize the events of our application. The swim lanes between the pivotal events do not necessarily line up and they do not need to, but we kept all our customer interactions in the top swim lane across the timeline. Our pivotal events have defined some boundaries, which help us see where a flow might be passed to another system or a new phase of the business. The swim lanes will break up the events for a phase into synchronous flows.

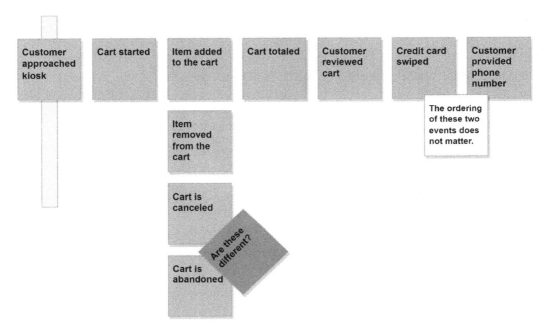

Figure 3.5 – A close-up of the cart flow

The events are organized within the swim lanes into horizontal and vertical flows. The horizontal flows represent the chronological flow of the events, and the vertical flows represent the alternative events that may create branches. Also shown in *Figure 3.5*, in the top-right corner, are two events that may occur in any order, but both would be expected to happen for the flow to continue. The two-dimensional nature of the timeline sometimes makes the placement of the notes and their relationship to other notes unclear. When you are in a similar situation, use a **comment** sticky note to clear up any potential misunderstandings. There is not any official legend for comment sticky notes, so use what works for your workshop by picking a color and size combination you won't have any conflicts with.

The two **cart is canceled** and **cart is abandoned** events could be different or the same, with unimportant semantic differences.

We have marked this question with a **hotspot** sticky note, and it can be addressed in the subsequent steps.

The flow for taking an order and then fulfilling it would appear to be the bulk of the application, and it certainly is for our simple application. Managing stores and bringing bots online have been placed on the left of the flow, as shown in *Figure 3.5*, which could happen at any time. At least one available store and one online bot must exist before we can manage taking an order from a customer.

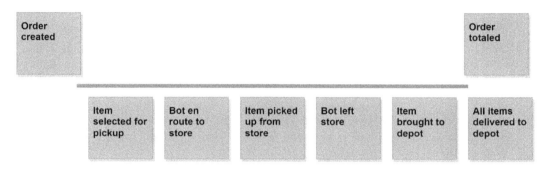

Figure 3.6 – Depot events in sync with order processing events

Additionally, we can see in *Figure 3.6* that a great deal occurs during the collection of order items in the depot, following the creation of an order and before it is touched again.

## Step four – people and systems

Now that the flows are becoming clearer and the sequences of events have been organized, we should add who or what will be triggering them. We should also add any external systems that will be involved in the flows we have created. This simple step is going to bring to the surface a lot of assumptions, and as a result, several new flows and events will be added. Triggering events that lead to other event flows will also be discovered and identified.

During our imaginary workshop, the following people and systems were identified.

## People

- Store owners are external users that operate stores in the mall where the *MallBots* service is active. They take care of their store details and the store inventory that is available to the service.

- Store administrators are internal users that curate the participating store in the service.

- Customers are external users who are visiting the mall and place orders to have items picked up from stores while they do other shopping.

- Bots are the AI processes that are running to control the robots that navigate the mall and pick up the items for an order.

- Depot administrators are internal users that manage the depot operations and monitor the robots.

- Depot staff are internal users working at the depot that are responsible for order fulfillment. This is a role, and the same people that do administration may also be doing fulfillment at the depot.

## Systems

- An SMS notification service is an external system responsible for contacting customers via SMS. We know that we will be sending text messages to users, but the specific service is not mentioned or decided at this time.

- A payment service is also an external service, which will be responsible for processing payments for the invoices associated with each order.

Thinking back to the previous chapter and the description of the types of bounded contexts, we can presume that the flows that use these external systems are going to be generic contexts. Neither payments nor notifications offers any competitive advantage to develop them just for our application.

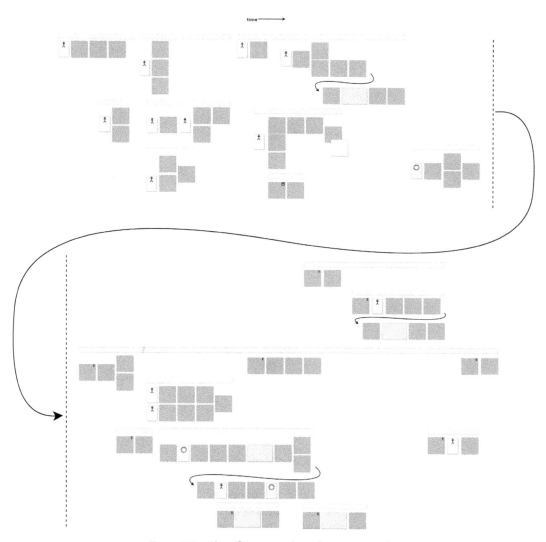

Figure 3.7 – Identifying people and systems results

The timeline is starting to look remarkably busy and wide. The participants have better organized the timeline into individual flow threads that visualize very well the number of events that are happening at any one moment.

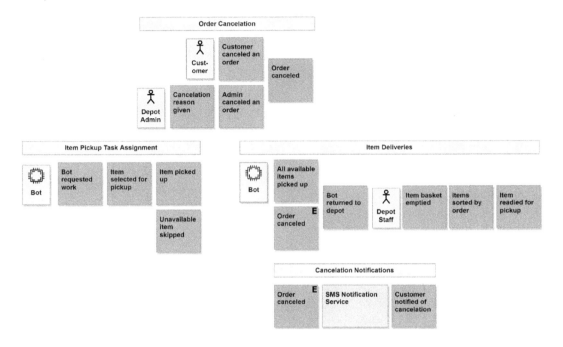

Figure 3.8 – Adding labels above event sequences

Now that the timeline has more events, actors, and services and has spread out to either side, it is a suitable time to label the flows we are defining about each group or event sequence, using some labeling tape of the tacky, removable variety. This would be an example of the fuzzy nature of EventStorming. This works for the participants and is not a requirement of this step or the workshop but labeling the timeline can help participants identify the business departments or operations more easily.

Another customization that the participants can add is markings to some of the events that trigger many of the flows.

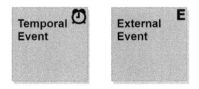

Figure 3.9 – Temporal and external variations of the domain event

These two events follow the same rules as the other events we have been putting up on the timeline in previous steps, but we have chosen to include a small marking in the top-right corner of the sticky note. A temporal event might use an alarm clock marking, as displayed in *Figure 3.9*. A drawing of an analog clock for events that happen at a specific time of day, as well as using a calendar for events that happen on specific days, weeks, or months, may be more meaningfully accurate than always displaying an alarm. External events are events that have happened elsewhere in the application that are then used to start up or kick off more work in other parts of it. All the external events are duplicates of their original event and are placed vertically in line with the original, or somewhere after it.

We have answered the question regarding who cancels orders from the previous step by showing that two people can act. The flow for item deliveries has a mix of actors and event triggers that might cause it to begin. The event that triggers the flow is an external event that comes from the previously mentioned **order cancelation** flow. It will not matter who cancels the order when we want to have the bot stop picking up items for an order and return to the **depot**. The same is true for the flow of **cancelation notification**, and if the customer canceled the order, they should still be notified of the fact.

In the previous step, there was an event at the end of the **item deliveries** flow that we no longer have here. For reference, this is what it looked like in the previous step:

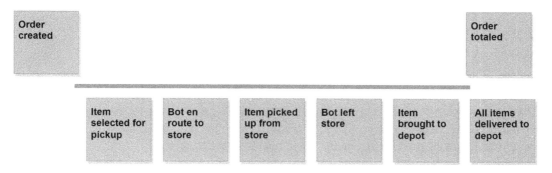

Figure 3.10 – A partial view of the depot and order flows from the enforcing the timeline step

In *Figure 3.10*, we have as part of the depot flow an event for **All items delivered to depot**, and now as shown in *Figure 3.8*, the **item deliveries** flow ends at **item readied for pickup**. The missing event has been rewritten and moved up into the overarching ordering process. When we applied context-specific ubiquitous language to the **item deliveries** flow, it became apparent that the depot is not responsible for knowing when all of the items have been collected for an **order**, at least not directly. What we need the depot to be capable of doing is processing the items as they come in and go out. Not to get ahead of ourselves, but it will likely be the responsibility of an order process manager.

There is the possibility that an item is not available at a store, so the bot would be unable to collect it. When this happens, the **order** should be updated so that we do not end up charging the customer for items we are unable to deliver. Overall, the entire flow associated with item pickup and deliveries will involve a substantial amount of thought and rework. There is nothing special about any event or sequence that would prohibit the participants from making corrections and improvements as the workshop advances through the steps.

### Step five – explicit walk-through

This is the step where we check our work by reading aloud the events as a story. Different participants will take turns walking through portions of the timeline. We do not want to have the events read verbatim from the timeline but to have the participant become a storyteller and narrate a story for the group, using the events as their outline. The point of reading the events as a story is that it will force the storyteller to think about how the events connect. When that becomes impossible or difficult, then we might have discovered a plot hole, or a missing or misplaced event. The audience participates in the process as well by pointing out the problems in the story.

This step will ask a lot from the participant doing the storytelling. Not only will they need to repeat portions of their story when corrections have been made but they will be interrupted constantly. They will need to add and move events or rewrite them when the ubiquitous language or narrative of their story is being lost. The facilitator can help with the events and changing the participant at pivotal events or flows can allow them to rest.

Storytelling will take a large amount of time to get through, and it could become the longest session of the workshop, so it is something to keep in mind when planning your session schedule. The storyteller has two tasks: the first is to tell their story aloud, and the second is to put one of their hands onto an event that becomes relevant to the story as it progresses. Combining these two tasks to reveal problems such as missing, out-of-place, or erroneous events is well worth the effort.

When telling the story, it is possible to overlook a significant event. During storytelling, we are connecting a target event with the ones that come after it, and we may not give any thought to the events that must occur directly before the target event. To do that, we can use an additional storytelling technique.

### Reverse narrative – storytelling in reverse

We focus on events flowing seamlessly into the next in storytelling, and this perspective may miss significant events. In reverse storytelling, the focus shifts to determining the event or events that directly precede an event. To start with using **reverse narrative**, we need to pick an event toward the end of the timeline and ask for the events that directly precede it. Discover any missing events and then repeat until you are at the earliest or leftmost events. Going over the timeline in reverse is also going to take a long time, so if you are in a rush, you might want to use the pivotal events and work backward from them or take a vote for the flow and event to work backward from.

## Storytelling results

As expected, the efforts of storytelling did an excellent job at uncovering and discovering implicit events hidden from us. The entire timeline was modified in one way or another, longer chains of events were discovered, flows were merged, and flows were dropped entirely.

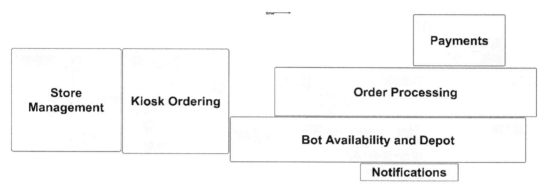

Figure 3.11 – Explicit walk-through storytelling groups

Knowing that this will be a lengthy session in real life, identifying groups in advance will benefit session participants in determining appropriate stopping points for breaks.

Continuing with our imaginary workshop example, we will break up the timeline into several groups and assign a different storyteller to each group.

## Store management stories

Starting from the left side of the timeline, we begin with stories about managing stores. From the story about creating a new store, there was no mention of also adding any products. You might be thinking the act of adding the products was implied in the details, but the purpose of this session is to make the implicit explicit, so we should add it.

If the storyteller does not notice an omission or other plot hole in their story, the participants should speak up to ask about what they think the problem might be. This process can also discover flows that are not as unique as originally thought when two stories sound too similar to one another.

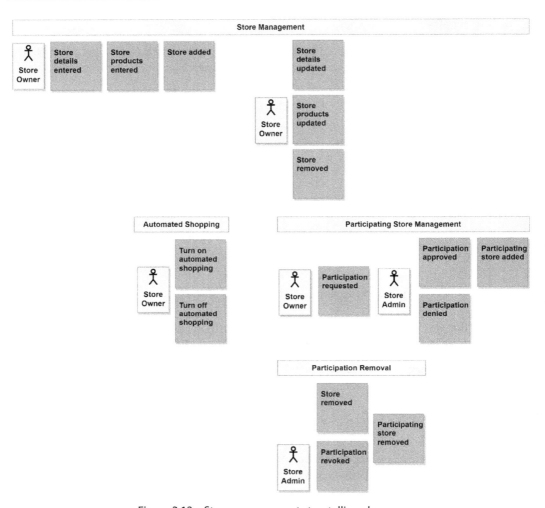

Figure 3.12 – Store management storytelling changes

Another issue was discovered while trying to tell the story about temporarily closing a store. Using the existing flows that we had, the stories would include adding a store, removing the store, and then adding the store back again. Any store with many products is not going to want to do that. We are not interested in keeping track of when a store is temporarily closed. It also might not be that the store is closed but that it wants to temporarily opt out of the automated shopping service. The solution was to add the flow for turning on and off the automated shopping feature for stores.

## Kiosk ordering stories

Moving right along into the next set of flows, we pick a new storyteller and begin. While the storyteller was telling the story involving adding items to the cart, someone interrupted to point out that the customer wasn't being given a new total. The same was also happening when the customer would remove items from the cart. To see the new total, they would have to restart the checkout process. To address this, changes to the events were made so that the customer would see the total after making any changes to the items in the cart.

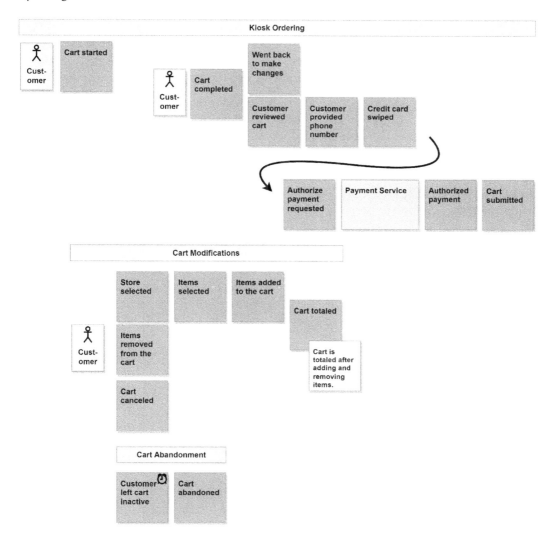

Figure 3.13 – A view of the kiosk ordering flows

> **Author notes on arrows**
>
> EventStorming should take place in a room that can provide a surface to work on an extremely wide timeline. On a real paper timeline, the use of arrows should absolutely be avoided, primarily because of the permanent nature of the marking. I am using arrows to compact the timeline horizontally and to connect parts of a single flow that I have split. This helps me fit the flows on the page, where I do not have the option to roll out more paper to fit wider timelines.
>
> Of course, workshops working with digital timelines should do what they want and relax the arrow rule.

Another implicit detail that was discovered was within the story of updating the cart items. To add items, we would need to have selected a store first. The **store selected** flow was expanded to include the adding of items, and a new **items removed from the cart** flow replaced **cart items updated**.

The last set of discoveries was in the checkout. There was no escape hatch. Once a customer had completed selecting the items, they were committed to that choice. A branch that covered the situation of a customer wanting to make changes was added and positioned where they could decide to proceed with their choices.

The **cart saved** event was removed from the timeline after the definitions of both a cart and an order were discussed. We accepted for a while that the submission was the transitory event between a cart and order, so saving a cart had no purpose. Removing the event helped make that clear to everybody.

Why we were expecting the customer to swipe the card wasn't clear either. It was to associate the card with the **order** so that it can be looked up later, and that seemed clear enough by looking at the timeline from the previous step, at least for the storyteller. It was also so we could authorize the card to avoid incidents of random passers-by creating orders they have no intention of ever picking up. This time, storytelling helped us discover an additional implicit reason or result for an event.

## Bot availability stories

A quick couple of stories uncovered some more implicit knowledge having to do with **bot availability**.

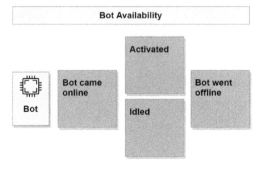

Figure 3.14 – A view of the bot availability flow

Determining whether a **bot** is available involves more than it being on or off. A **bot** is available when it is on and idle and it is unavailable any other time. The term available in the context of the depot and the bots was also defined and added to the ubiquitous language:

- **Bot availability**: A bot's readiness to be given work

## Order processing stories

Next, we turn to the rather large ordering process, and we dive first into the **order creation** flow. There is a change to be made to reflect the removal of the **cart saved** event, and we swap it out for **cart submitted**. In the telling of our stories, we make an immediate jump from receiving the cart submission to checking an order for problems. Checking the order also happens before we create it later in the flow. The flow needs to start with events dealing with the cart and then transition to events for the order only after we are finished with the cart. This is how the updated flow looks:

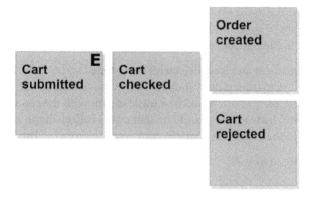

Figure 3.15 – A view of the order creation flow

An audience member interrupts the story of a customer canceling their order to point out that the story is missing a beginning. The beginning that we were missing needed to answer, "*How does the customer get to the order to be able to cancel it?*" The kiosks are the only interfaces that a customer could use, so they would need to visit one of those. To find their order, they would use the credit card associated with the order and swipe it in the card reader. We make the changes to fix the story, and the results are shown in *Figure 3.16*:

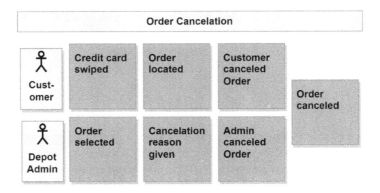

Figure 3.16 – A view of the order cancelation flow

Perfect! We have the beginning to the story and the customer is no longer magically canceling orders with their mind. It felt right to do a similar update for the admin cancelation flow, and the **order selected** event was added to the beginning to make the selection of an order explicit.

The **item pickup task assignment** and **item deliveries** flows ended up being completely removed and replaced as the result of our storytelling. Bots wouldn't be sent out to pick up individual items and then return to the depot. The singular pickups would be a waste of time with the amount of back-and-forth trips a bot would be making. Instead, they would be sent out to collect all the items for an order and then return after visiting all the stores necessary to complete the shopping list.

A new definition was recorded as well:

- **Shopping list**: A list assigned to bots containing the stores to visit and the items to pick up

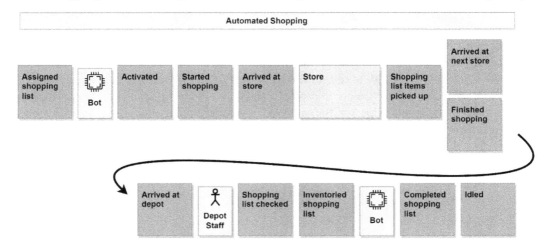

Figure 3.17 – A view of the automated shopping flow

As seen in *Figure 3.17*, the perspective of how the bots would receive their work has also changed. They would not be in a position where they would poll or request work, but instead, we would rely on their availability status to assign them work. A very important external system, the **store**, was included in the new combined flow. Determining if all available items have been picked up was removed as a responsibility of the **bot** and moved later in the flow, where it was given to the **depot staff**.

### Invoicing stories

The last major changes would be made to the **invoicing** flows. First, we updated the flows to reflect the changes made previously in the **kiosk ordering** flow, specifically the usage of a pre-authorization for the customer's credit card.

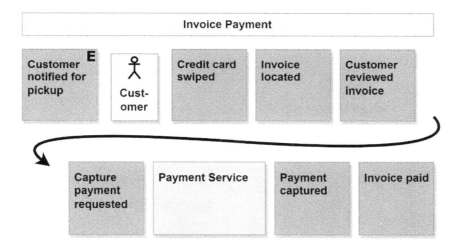

Figure 3.18 – A view of the invoice payment flow

The customer was also allowed to review the invoice they were being expected to pay, and the **customer reviewed invoice** event was added to reflect that.

## Step six – problems and opportunities

With any luck, the last session opened the participant's eyes to what it is we intend to be building, and we have mostly started agreeing on how it will get done. Of course, by going through the whole timeline and, in some cases, specific portions at a deeper level, we are sure to dig up even more questions and ideas.

We close out our Big Picture workshop with a short session, where we ask everyone to place hotspots where they think problems still exist and to place opportunity stickies up where they have ideas for improvements.

Throughout the workshop, we focused on the current goal or version of the idea, so our problems should focus on issues that would exist for that version. The opportunities will be focused on the next version and beyond.

## Identifying the contexts

We can place boxes around the various groupings we have on the timeline to identify the bounded contexts we've discovered:

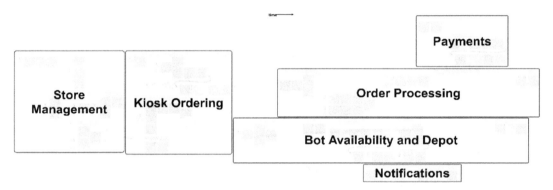

Figure 3.19 – The bounded contexts of MallBots

The sections we identified earlier could also work as our bounded contexts. This won't always be the case of course. Determining bounded contexts is an art and not a science, but we shouldn't base them on how much a storyteller may be willing to narrate.

## Wrapping up

That was a lot to go through. Had this been a real EventStorming workshop, every participant should feel good about the final result. What couldn't be shown in the previous sections was the amount of movement and rearranging that can take place during any given session. The discovery process will redo portions of the timeline multiple times. It is better to throw away a sticky note that is no longer needed than to get the solution to the business problem wrong in the code, which could be more costly to redevelop.

## Design-level EventStorming

We can now go further with a design-level workshop. This format looks at a **single core bounded context**, and we won't need to invite the domain experts and developers that would be relevant to the context in question. We want to add to our Big Picture events the other concepts that turn a flow of events into a process, which follows the EventStorming flow shown in *Figure 3.1*. The goal of the workshop also changes. In the Big Picture workshop, we were focused on exploration and discovery; now, we want to do design and process building.

### Picking the right contexts to focus on and design

DDD will not be applied to every part of your application. We can tell from our efforts with the Big Picture workshop that there are several parts of our application that we may not want to invest time in to dig into deeply or model the context using tactical DDD patterns. **Store management** could be a Supporting context that we can use entities with simple **Create, Read, Update, and Delete (CRUD)** access patterns. The **payments** and **notification** contexts are expected to be external services and won't need any special attention either. That leaves the **depot** and **order processing** contexts. These are the contexts that are complex enough to be Core contexts and warrant a design-level workshop, giving us the best chance of getting them right from the start.

# Understanding the business

We now arrive at the space between *designing our application* and *planning how that will happen*. Leaping over this gap right into planning might mean we lose some of the hard work that a lot of people helped put together. We need something that bridges this divide that can capture the knowledge that has been shared with us and can be used to test us to keep us honest. For this, we turn to executable specifications and **Behavior-Driven Development (BDD)**.

BDD is a form of living documentation that, in most cases, can be formatted in a way that makes it machine-readable, so it can be used as part of a **continuous integration and continuous delivery (CI/CD)** pipeline to perform acceptance testing – all while still being completely readable by non-developers. The purpose of the documents that we create using BDD is to keep the distance between what the business needs are and what is developed to implement that need as small as possible. Domain experts and developers will share and collaborate on the documents.

If you are already doing EventStorming or using some other DDD tool to develop the ubiquitous languages and the bounded contexts, then you are going to be able to ease into BDD. We will take the capabilities of each bounded context, break those down into features, and then provide example scenarios for each feature.

The scenarios should be written in such a way that they describe what we want the application to do and not how we want it to be done. For example, if you were writing scenarios for an authentication module, the following would be a poor example of a scenario:

```
Feature: Authenticate Users

  Scenario: Login to the application
    Given a user with username "alice" and password
      "itsasecret"
    When I enter the username "alice"
    And I enter the password "itsasecret"
    And I click the "Login" button
    Then I see the application dashboard
```

This scenario focuses too much on how a user is authenticated. The user must enter a username, enter a password, and then click a button. What if authentication uses a fingerprint scanner, or a smartcard authentication method instead? This scenario would not work with alternative forms of authentication; we need a better scenario that focuses on how authentication should work:

```
Feature: Authenticate Users

  Scenario: Login to the application
    Given an active user "alice"
    When "alice" authenticates correctly
    Then "alice" can access the application dashboard
```

We will use Gherkin (https://cucumber.io/docs/gherkin/reference/) to write our features and use the Godog tool (https://github.com/cucumber/godog) to execute them as our acceptance tests. This is what our specifications will look like:

```
Feature: Creating Stores

  As a store owner
  I should be able to create new stores

  Scenario: Creating a store called "Waldorf Books"
    Given a valid store owner is logged in
    And no store called "Waldorf Books" exists
    When I create the store called "Waldorf Books"
    Then a store called "Waldorf Books" exists
```

This is a Gherkin-formatted feature that could be written by a domain expert:

1.  The first line sets the `Create Store` feature name and is required.

2.  The next two lines are the user story, which sets our expectations but is optional.

3.  The rest of the file is the scenario for creating a store called `Waldorf Books`, which could be followed by more scenarios to provide further examples.

We leave out the details of how creating a store should be done. There are no mentions of saving records, performing search queries, or any references to specific user interface details.

On its own, it is documentation providing a feature story and an example scenario. We can place this into our repository at `/stores/features/create_store.feature` and then implement the feature and scenario using Go in `/stores/stores_test.go` with the following:

```go
var storeName = ""

func aStoreExists(name string) error {
    if storeName != name {
        return fmt.Errorf(
            "store does not exist: %s", name,
        )
    }
    return nil
}

func aValidStoreOwner() error {
    return nil
}

func iCreateTheStore(name string) error {
    storeName = name
    return nil
}

func noStoreExists(name string) error {
    if storeName == name {
        return fmt.Errorf("store does exist: %s", name)
    }
    return nil
```

```
}

func InitializeScenario(ctx *godog.ScenarioContext) {
    ctx.Step(
        `^a store called "([^"]*)" exists$`,
        aStoreExists,
    )
    ctx.Step(
        `^a valid store owner is logged in$`,
        aValidStoreOwner,
    )
    ctx.Step(
        `^I create the store called "([^"]*)"$`,
        iCreateTheStore,
    )
    ctx.Step(
        `^no store called "([^"]*)" exists$`,
        noStoreExists,
    )
}
```

This test implementation does not do much other than demonstrate the code necessary to turn the feature documentation into executable specifications, which can be executed to validate that the developers have implemented the feature as specified.

We can use the results from the Big Picture workshop and the stories to write the features that will be implemented as we develop more and more of the *MallBots* application. The Gherkin feature files will be available in the code repository, and implementations will be added in advance throughout the development of the application.

## Recording architectural decisions

Moving to the implementation side, we now face decision-making issues on how this application will be developed. The decisions that we make will have lasting repercussions for a project, and over the life of the project, the motivations for why a decision was made can be lost.

Enter the architecture decision record. The most popular format for these records comes from Michael Nygard (https://cognitect.com/blog/2011/11/15/documenting-architecture-decisions.html) who suggested the following format, shown here in Markdown:

```
# {RecordNum}. {Title}

## Context
What is the issue that we're seeing that is motivating this
decision or change?

## Decision
What is the change that we're proposing and/or doing?

## Status
Proposed, Accepted, Rejected, Superseded, Deprecated

## Consequences
What becomes easier or more difficult to do because of this
change
```

An architectural decision record should be made each time a significant change is being made to software, the infrastructure, or the service dependencies.

Here are some examples of decisions you might want to record:

- Choosing to use a cloud provider
- Adding or replacing infrastructure to address performance or availability issues or concerns
- Going with a non-standard solution for a particular reason
- Deciding on a new programming language for new or revised code
- Adopting DDD or other design patterns

The popularity of the preceding format is thanks in part to its simplicity. The small template is quick to fill in, commit, and start the decision conversations. Additional headings may be added to provide even more information when necessary.

Another point in favor of ADRs, in general, is that they are kept in the code repository, making them very easy to find. They are treated like an immutable log, and when new decisions are made, a new document is written to support them. The exception to that is when an older decision is being replaced, and you will need to update its status to reflect that.

Our architecture decision log is kicked off with two decisions:

- **Keep an architecture decision log**: The decision on whether to keep the record of decisions, which will have significant impacts on the application

- **Use a modular monolith architecture**: We want to avoid a mess with an unstructured monolith and the deployment complexities involved with developing an application with microservices

This will be enough to get the ball rolling, and we will see new decisions being made in each new chapter.

## Summary

In this chapter, you read your way through an experience of the Big Picture workshop to dig up all the implicit details and knowledge locked away in the minds of the make-believe domain experts. We also covered executable specifications and will be hearing more about them as we complete each new portion of the application. We were also introduced to ADRs and used them to log our first two decisions. This design and planning chapter concludes the first section of the book.

In the next chapter, *Event Foundations*, we will be developing the application using event-driven architectures. In the next chapter, we will make use of our first architecture decisions and lay the foundation for our modular monolith application.

## Further reading

- *Introducing EventStorming*, by Alberto Brandolini, available at `https://leanpub.com/introducing_eventstorming`

- *Awesome EventStorming*: `https://github.com/mariuszgil/awesome-eventstorming`

- *Awesome BDD*: `https://github.com/omergulen/awesome-bdd`

# Part 2: Components of Event-Driven Architecture

In this part, we will begin and complete a journey of transforming MallBots from a synchronous application into an asynchronous application. We will introduce and refactor the application using domain events, event sourcing, and messaging. This part will also provide an introduction to and hands-on experience of dealing with eventual consistency and other complications that developers must contend with when developing event-driven applications.

This part consists of the following chapters:

- *Chapter 4, Event Foundations*
- *Chapter 5, Tracking Changes with Event Sourcing*
- *Chapter 6, Asynchronous Connections*
- *Chapter 7, Event-Carried State Transfer*
- *Chapter 8, Message Workflows*
- *Chapter 9, Transactional Messaging*

# 4
# Event Foundations

In the first part of this book, we discussed what event-driven architectures are and the other patterns we might use when developing them. We then dove into the design and planning of an application, and we'll be implementing event-based approaches to the existing synchronous methods it uses now. This next part will introduce you to event usage, tracking, and forms of communication, and will also refactor the *MallBots* application into a fully event-driven application. Each chapter will cover a different pattern and accompanying implementation, which will build on what was learned in the previous chapters.

In this chapter, we will take a look at how the application is being built and how the modules of the application communicate. After a tour of the application, we will refactor portions of the application to use domain events, a domain-driven design pattern, to set the stage for our future refactoring efforts.

We will work with the following main topics:

- An in-depth tour of our monolithic application structure and design
- A look at the synchronous integrations of the application we are working with
- An introduction to the types of events we will be using
- Implementing domain events to refactor how side effects are handled

## Technical requirements

In this chapter, we will be implementing domain events for our application. You will need to install or have installed the following software to run the application or to try the examples:

- The Go programming language – version 1.17+
- Docker

The source code for the version of the application used in this chapter can be found at https://github.com/PacktPublishing/Event-Driven-Architecture-in-Golang/tree/main/Chapter04.

# A tour of MallBots

Our *MallBots* application is a modular monolith, which, if you recall from *Chapter 2, Supporting Patterns in Brief*, is an application design that sits somewhere between a classic monolith design and a microservices application design. We have most of the benefits of both designs with only a few downsides.

## The responsibilities of the monolith

The root directory of our code is kept minimal and what stands out is the module names. We intentionally avoid the use of generic or general layer names, such as *controllers*, *config*, or *models*, in the root directory. We use application component names, such as *baskets*, *stores*, *depot*, and *ordering* instead, so that we end up with a code repository that looks like an application that deals with shopping and not like some generic, no-idea-what-it-does application. Each of these modules is a different bounded context in our application.

> **Screaming architecture**
>
> The organization we're using for our root level directory structure is called **screaming architecture**, credited to *Robert C. Martin*, and more details can be found in this 2011 post: `https://blog.cleancoder.com/uncle-bob/2011/09/30/Screaming-Architecture.html`.

It isn't just modules that we will find in our root. We do have some other directories that are necessary:

- `/cmd`: A typical root-level directory for Go applications. This directory will contain one or more child directories for each eventual application we can generate from the code.

- `/internal`: A special directory for Go code bases that receives special treatment by the compiler. Anything in this directory or its child directories is only accessible to the parent directories or sibling directories of the internal directory. Seeing an internal directory is a signal to other developers that the code within is not meant to be imported into any outside applications, and this intention is backed up by the compiler.

- `/docs` and `/docker`: Additional utility directories containing documentation and scripts to aid in the understanding and local development of the application.

### *Shared infrastructure*

The monolith creates or connects to all parts of the infrastructure that it and the modules will be using. References to the infrastructure are then passed into each module to be used in whichever way that module needs to use the connections. The modules will not create any connections to the infrastructure themselves:

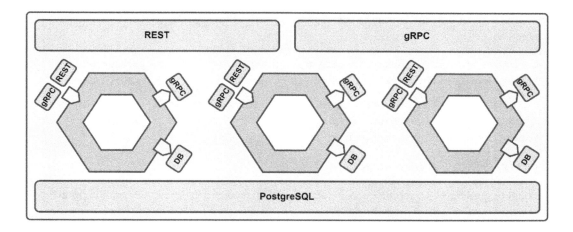

Figure 4.1 – The monolith and module infrastructure

*Figure 4.1* shows the current state of the application we will be working with. Around the center hexagons, we have the monolith and the infrastructure such as the database and the exposed APIs. Aside from instantiating the modules and supplying them with the dependencies that they need, we can keep the monolith code very simple. The modules, represented by the hexagons in the middle, are each initialized with references or connections to the infrastructure.

## Module code organization

Each of the modules that makes up our application exposes a protocol buffer API and a small module file that contains the composition root for the module code. The modules also have their internal packages to keep unintentional imports from being made between the modules.

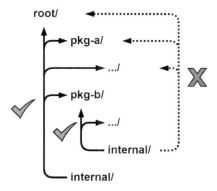

Figure 4.2 – Internal package import rules

*Figure 4.2* illustrates how the multiple internal packages help us manage our relationships and control the dependencies between the modules:

- /root/internal: This package can be imported by /root and any package found in the directory tree under it.

- /root/pkg-b/internal: This package may only be imported by /root/pkg-b and any package found in the directory tree under it. Both /root and /root/pkg-a will not be permitted to use any imports from this package.

### Accept interfaces, return structs

The idiom or guideline "*Accept interfaces, return structs*", first coined by *Jack Lindamood* in his article *Preemptive Interface Anti-Pattern in Go* (https://medium.com/@cep21/preemptive-interface-anti-pattern-in-go-54c18ac0668a), works very well with ports and adapters or hexagonal architecture. This guideline is followed by all modules, even simpler ones such as the **Payments** and **Notifications** modules.

However, this guideline and ports and adapters are not quite the same, as the intentions or goals are slightly different. When you follow the guideline as it is set out in the article, you have the consumer of the concrete value define the interface it requires, as shown in this example:

```
// in db/products.go
type ProductRepository struct {}
func NewProductRepository() *ProductRepository {}
func (r ProductRepository) Find() error {}
func (r ProductRepository) Save() error {}
func (r ProductRepository) Update() error {}
func (r ProductRepository) Delete() error {}
// elsewhere in services.go
type ProductFinder interface {
    Find() error
}
func NewService(finder ProductFinder) *Service { }
```

In the preceding code, NewService will accept anything that implements the ProductFinder interface. The interface definition is kept close to the consumer, ideally in the same package or file. It is also defined to be as small as possible, only requiring the methods that the consumer would need to use. Smaller interfaces lead to more freedom in what concrete values you may be able to accept. In this situation, both the interface and implementation are loosely coupled, and the maintainer of ProductRepository may be unaware that ProductFinder exists.

On the other hand, when working with ports and adapters, we want to define contracts for the interactions with our applications. This often means we will be defining larger interfaces that work as the contracts for the application adapter implementations. These interfaces will not sit next to each consumer but will be kept in a central location, such as in the application or domain directory. The reverse is also going to be true for the maintainers of the implementations. The implementations will be written after the interfaces, and will be built to satisfy one or more interfaces.

Using interfaces will result in easier to test code, so teams should use the approach that fits the situation.

---

**Interface checks**

Most implementations written to a contract interface will be used somewhere that has a static conversion and would be caught by the compiler – for example, `*os.File` used in a method accepting `io.Reader`. When there are no static conversions that the compiler can catch, then a change to the implementation may break that contract but won't keep the application from being compiled. It won't be until the application is running that you may notice the issue. A solution to this problem is to add an interface check that the compiler can catch but that will then be left out of the built application:

```
var _ TheContractInterface = (*TheContractImplementation)(nil)
```

Here, we create a `TheContractImplementation` value that is assigned to `_` with the `TheContractInterface` type. This adds a static conversion, and we can trust that any issues in our implementation will now be caught at compile time and not left to be discovered by the user after deployment. The assigned value is never used and will be excluded from the compiled output for our application.

---

Using interface checks, and placing them next to implementations meant to satisfy any given interface, will protect you in the rare occurrence that there isn't a static conversion elsewhere in the application.

## Composition root

The internal design of each module may differ, but they all use the same pattern to start up. A composition root is the part of an application where you bring the infrastructure, configuration, and application components together:

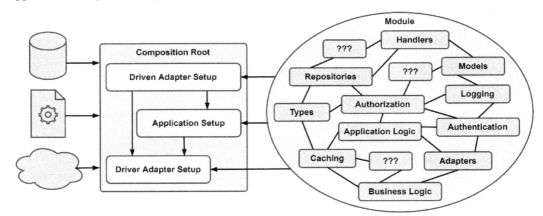

Figure 4.3 – Using a composition root to build application dependencies

The composition root is also where dependency injection takes place, and an application object graph will be constructed. For our modules, we will undertake the following actions:

- Construct the Driven adapters
- Construct the application and inject the Driven adapters
- Construct the Driver adapters and inject the application and Driven adapters

This snippet from the Notifications module shows these three steps in action:

```
// setup Driven adapters
conn, err := grpc.Dial(ctx, mono.Config().Rpc.Address())
if err != nil { return err }
customers := grpc.NewCustomerRepository(conn)
// setup application
var app application.App
app = application.New(customers)
app = logging.LogApplicationAccess(app, mono.Logger())
```

```
// setup Driver adapters
grpc.RegisterServer(ctx, app, mono.RPC())
```

1. The **Driven** adapters implement the ports in the **application** and only need infrastructure to be constructed.

2. The **application** is constructed next and needs the **Driven** adapters but not the **Driver** adapters.

3. Finally, the **Driver** adapters are constructed using a combination of infrastructure and the **application**. At this level, we are more concerned with concrete values and try to avoid abstractions. This pattern is simple, predictable, and boring, and all three of those characteristics are positives.

---

### Dependency injection tooling

Composition roots are nothing more than lines of code creating instances that are then used in the construction of more instances, ultimately building a dependency graph. There are tools for Go that can be used to do this task, such as **Google Wire** (`https://github.com/google/wire`), which uses code generation to build the wiring between the dependencies. Another tool, Dig (`https://github.com/uber-go/dig`), is a runtime dependency tool that uses reflection. Deciding to use a tool versus maintaining the code yourself is not without trade-offs. Using some tool to manage the dependencies and build the graph is not worth the effort until the number of dependencies or the complexity of the graph has grown too large to keep straight.

---

## Protocol buffers and the gRPC API

The communication between the modules is entirely synchronous and uses protocol buffers and gRPC. Each module that has exposed a gRPC service API will share it from a package with the following naming structure: `/<module>/<module>pb`. For example, `/stores/storespb` would be where to find the gRPC service API for the **Store Management** module. The gRPC service APIs are outside of the module's internal package, and it is all that is exposed for other modules to use.

---

### Buf

We will be using `buf`, `https://buf.build/`, a tool to compile our protocol buffer files into Go code. The primary advantage of using this tool instead of the `protoc` compiler directly is the ability to manage the complexity of the compilation rules by using configuration files. We are also able to enforce a coding standard for the gRPC APIs and message structures using the linting features built into the tool.

---

We could use any other synchronous method to connect the modules that doesn't result in a cyclic dependency and a compile-time error. With help from our composition root and dependency injection, we avoid this problem and can have two modules depending on each other. This is the case for the **Ordering** and **Payments** modules; they each make calls to the other.

A single gRPC server may serve any number of gRPC services if the compiled protocol buffers do not have any namespace conflicts. To avoid this conflict, we make sure to compile the parent directory name as part of each protocol buffer API. We end up with `basketspb.Item` and `orderingpb.Item` and avoid all conflicts.

## User interface

There is a REST API for users to use that comes from the modules exposing their gRPC service APIs using `grpc-gateway` (`https://github.com/grpc-ecosystem/grpc-gateway`). Most modules expose most of their gRPC services this way; notable exceptions are the **Notifications** module and most of the **Depot** module.

The REST APIs are mounted at `http://localhost:8080/api/*`.

### *Swagger UI*

To make things easier to experiment and run examples, the REST APIs can be accessed with the Swagger UI found at the root of the web server: `http://localhost:8080/`.

## Running the monolith

The monolith and the process it depends on can be started using Docker Compose. Navigate to the root of the chapter and run the following command:

```
> docker-compose up -d
```

After a short time downloading the required containers and compiling the monolith, you'll be presented with the command prompt again, and you should run the following:

```
> docker-compose logs -f
```

You should see the logs from the Postgres and monolith containers, and that output should look something like the following:

```
postgres     | LOG:  database system is ready to accept
connections
monolith     | started mallbots application
monolith     | web server started
monolith     | rpc server started
```

The order in which the containers logs are reported may be different, but if we see that the database is ready for connections and the monolith and its servers have started, then we are good to go.

> **Stopping and Rebuilding the Containers**
>
> Use *Ctrl* + *C* to exit the logs command. Then, use `docker-compose` down to stop and remove the containers. If you make any changes to the monolith code, you will need to append `--build` to the `compose` up command to recompile and rebuild the container.
>
> For more information on using `docker-compose`, visit `https://docs.docker.com/compose/reference/`.

Open your browser and visit `http://localhost:8080/`. What you should see now is the Swagger UI. In the top-right corner will be a dropdown where you can access the REST APIs for the different modules:

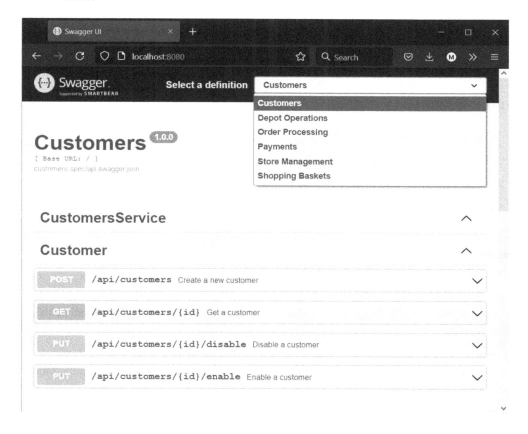

Figure 4.4 – The MallBots Swagger UI

You can use this UI to simulate the experience a store manager may have with the **Store Management** module or to experience creating an order, starting with a basket.

## A focus on event-driven integration and communication patterns

When you are taking your own look around the code repository, keep in mind that this application is put together to demonstrate how distributed components integrate and communicate with each other. Business rules and logic will be light, and in some places, there might be some digital handwaving at play, and implementations left unimplemented.

# Taking a closer look at module integration

As I previously stated, all interactions between the modules are entirely synchronous and communicate via gRPC. With a distributed system such as our modular monolith application, there are two reasons that bounded contexts will need to integrate:

- They need data that exists in another bounded context
- They need another bounded context to perform an action

## Using external data

When a bounded context needs data belonging to another bounded context, it has three options:

- Share the database where the data is stored
- Push the data from the owner to all interested components
- Pull the data from the owner when it is needed

The first option should be avoided in most situations, especially if changes are being made from more than one location. Rules surrounding invariants may not be implemented correctly or at all in every location.

When you push data out, you will be sending it to a list of known interested components. This is a maintenance nightmare. The bigger the number of components grows, the harder it will be to keep these lists correct.

Pulling data avoids having to deal with maintaining a list, but the trade-off is there will be more calls and a greater load put on the component that owns the data. Caching the data can help, but that inevitably leads to issues with invalidating stale cache data.

> **Tip**
> Given the options, pulling data is the better choice in most cases. The local component can be written to be ready for failures with retry logic, circuit breakers, and other mechanisms.

## Adding items to a basket

An `AddItem` request contains a product identifier and a value for the quantity of items to add. To complete the request, the **Shopping Baskets** module will need additional information for both the product being added and the store it is sold from. This information is pulled from the **Store Management** module the moment it is needed. The following logs show the calls made during an `AddItem` request:

```
monolith      | INF --> Baskets.AddItem
monolith      | INF --> Stores.GetProduct
monolith      | INF <-- Stores.GetProduct
monolith      | INF --> Stores.GetStore
monolith      | INF <-- Stores.GetStore
monolith      | INF <-- Baskets.AddItem
```

It is easier to pull data, and that is why it was done this way. **Shopping Baskets** is also not the only module that uses **Store Management** data. The modules that need product and store data could use different options. Some modules might have the data pushed to them and others might pull the data down. Deciding which option to use for the external data that a module uses is an *it depends* or a *case-by-case* situation.

> **Just the Basics Logging**
> The arrows in the log messages signify the entering and exiting of application methods. If a method has encountered an error, then a message in red would also appear on the exit rows.

## Commanding external components

To get a command to the bounded context that needs it, two options come to mind:

- Push the command to the commanded component
- Poll for new commands from the commanding component

The first option is going to be the most widely used. It is simple and straightforward and only needs an API endpoint to exist somewhere for the command to be sent to. The second option is more complicated to set up and may result in more calls and loads existing when there are no new commands to begin working through.

### Checking out a basket

In the current version of the application, when a customer chooses to check out their basket, the CheckoutBasket handler makes a single call into the **Ordering** module to create a new order. The CreateOrder handler, however, makes several calls to other modules, as shown in the following logs:

```
monolith    | INF --> Baskets.CheckoutBasket
monolith    | INF --> Ordering.CreateOrder
monolith    | INF --> Customers.AuthorizeCustomer
monolith    | INF <-- Customers.AuthorizeCustomer
monolith    | INF --> Payments.ConfirmPayment
monolith    | INF <-- Payments.ConfirmPayment
monolith    | INF --> Depot.CreateShoppingList
monolith    | INF --> Stores.GetStore
monolith    | INF <-- Stores.GetStore
monolith    | INF --> Stores.GetProduct
monolith    | INF <-- Stores.GetProduct
monolith    | INF <-- Depot.CreateShoppingList
monolith    | INF --> Notifications.NotifyOrderCreated
monolith    | INF <-- Notifications.NotifyOrderCreated
monolith    | INF <-- Ordering.CreateOrder
monolith    | INF <-- Baskets.CheckoutBasket
```

This is the most extensive call in the application and serves as an extreme example of a synchronous call chain. A total of seven modules are involved in the process of checking out a basket. This is, of course, for demonstration purposes, but call chains such as these can develop in real applications that rely on synchronous communication between components.

# Types of events

Let's cover the kinds of events we will be learning about and using along the journey to develop a fully event-driven application by the end of the book.

In an event-driven application and even in an application that is not event-driven, you will encounter several different kinds of events:

- Domain events – synchronous events that come from domain-driven design
- Event sourcing events – serialized events that record state changes for an aggregate
- Integration events – events that exchange state with other components of an application

## Domain events

A domain event is a concept that comes from domain-driven design. It is used to inform other parts of a bounded context about state changes. The events can be handled asynchronously but will most often be handled synchronously within the same process that spawned them.

We will be learning about domain events in the next section, *Refactoring side effects with domain events*.

## Event sourcing events

An event sourcing event is one that shares a lot in common with a domain event. These events will need to be serialized into a format so that they can be stored in event streams. Whereas domain events are only accessible during the duration of the current process, these events are retained for as long as they are needed. Event sourcing events also belong to an aggregate and will be accompanied by metadata containing the identity of the aggregate and when the event occurred.

We will be learning about and implementing these events in *Chapter 5, Tracking Changes with Event Sourcing*.

## Integration events

An integration event is one that is used to communicate state changes across context boundaries. Like the event sourcing event, it too is serialized into a format that allows it to be shared with other modules and applications. Consumers of these events will need access to information on how to deserialize to use the event at their end. Integration events are strictly asynchronous and use an event broker to decouple the event producer from the event consumers.

We will be learning about integration events in *Chapter 6, Asynchronous Connections*, and we will then see the different ways they are used in subsequent chapters.

# Refactoring side effects with domain events

We've talked about domain events before and spent a great deal of time thinking about them in the *EventStorming* exercise in the previous chapter. To refresh your memory, a domain event is a domain-driven design pattern that encapsulates a change in the system that is important to the domain experts. When important events happen in our system, they are often accompanied by rules or side effects. We may have a rule that when the `OrderCreated` event happens in our system, we send a notification to the customer.

If we put this rule into the handler for `CreateOrder` so that the notification happens implicitly, it might look something like this:

```
// orderCreation
if err = h.orders.Save(ctx, order); err != nil {
    return errors.Wrap(err, "order creation")
}
// notifyCustomer
if err = h.notifications.NotifyOrderCreated(
    ctx, order.ID, order.CustomerID,
); err != nil {
    return errors.Wrap(err, "customer notification")
}
```

If it were to remain as just one rule, we may be fine doing it this way. However, real-world applications rarely stay simple or have simple rules. Later, in the life of the application, we want to add a **Rewards** module to our application, we add the code for the rule to the same handler, and later, still we want more side effects to occur. What we had before, `CreateOrder`, should now be renamed `CreateOrderAndNotifyAndReward...`; otherwise, it won't properly reflect its responsibility. Also, consider there will be other rules and other handlers that may be implemented, so finding the implementations for a rule may become a problem.

Domain events will allow us to explicitly handle system changes, decoupling the work of handling the event from the point it was raised. Continuing with the previous example, our system would raise an `OrderCreated` event, and other parts may react to it to handle each rule that should follow it. The system I am speaking of is going to be a single bounded context, and the raising and handling of the event will be entirely synchronous and in-process.

To add domain events to the application, we will be implementing the following new features:

- Aggregates to raise the domain events
- Domain events to share state changes
- Dispatchers to handle the publishing of events that our rule handlers are subscribed to
- The plumbing to bring it all together

Here is a look at what the process to handle the side effect of sending a notification to a customer will be like after we are finished:

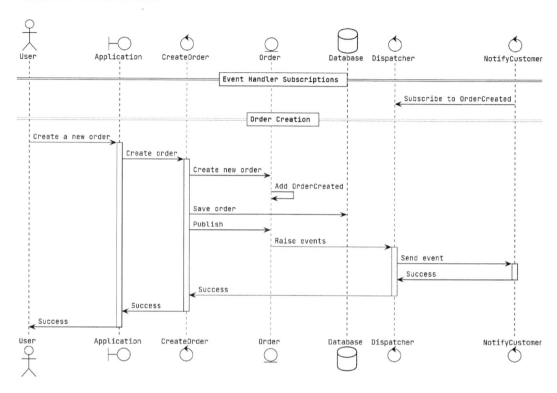

Figure 4.5 – Order creation with domain events

This is what the `Order` aggregate in the **Ordering** module looks like right now:

```
type Order struct {
    ID         string
    CustomerID string
    PaymentID  string
    InvoiceID  string
    ShoppingID string
    Items      []*Item
    Status     OrderStatus
}
```

We could add a slice for events with `[] Event` and the methods to manage them, but we know better, and there are going to be other aggregates and handlers we will be updating. To add the necessary event handling bits, we will make use of composition, and we'll end up with the following:

```
type Order struct {
    ddd.AggregateBase
    CustomerID string
    PaymentID  string
    InvoiceID  string
    ShoppingID string
    Items      [] *Item
    Status     OrderStatus
}
```

We added `AggregateBase` from the `internal/ddd` package and removed the `ID` field because that field now comes provided by `AggregateBase`. A small change to the couple of places we instantiate a new `Order` will also be necessary:

```
order := &Order{
    AggregateBase: ddd.AggregateBase{
        ID: id,
    },
    // ...
}
```

A quick word on the topic of field visibility in this application. You may have noticed in the recent code snippets that all the fields of our `Order` domain aggregate are public. I have chosen to use public fields, even though I know that means someone could make a change to the aggregate without using a method or domain function. Go does not have getters or setters, so you would need to create them yourself with something like the following:

```
type Order struct {
    id string
    // ...
}
func (o Order) ID() string        { return o.id }
func (o *Order) SetID(id string) { o.id = id }
...
```

This may not seem like a great deal for a single field example, but with a lot of models and even more fields, it does add up. If you decide to be very strict, then you would not implement any of the getters and would instead need to use builders and factories to construct the aggregates. In this application, I am choosing not to use private fields, but I am also not making the suggestion that this is the correct choice. Give both a go and decide for yourself.

The new `AggregateBase` and the interfaces it implements are straightforward:

Figure 4.6 – AggregateBase and its interfaces

The `Aggregate` interface also includes the `Entity` interface, which has a single `GetID` method defined. We will need this getter when we are working in methods that accept either `Aggregate` or `Entity`, to avoid having to determine the type of object we are working with to access the `ID` field.

Also straightforward is the first event we are working with, the `OrderCreated` event:

```
type OrderCreated struct {
    Order *Order
}
func (OrderCreated) EventName() string {
    return "ordering.OrderCreated"
}
```

Normally, we would want to have only the information we deem important in an event and take efforts to trim that down even further, but this is a domain event. Domain events will not be shared outside of the bounded context, module, or microservice they belong to. This means several things:

- We can put whatever we want into them so long as we are still treating them as immutable carriers of state

- We will not need to be concerned with anyone subscribing to them without knowing; therefore, there is no risk of making changes to them and breaking things unintentionally

- They live a very short amount of time and do not need to be serialized or versioned to be stored in any database or stream

The `OrderCreated` event has an `EventName` method that serves two purposes. The first is to satisfy the `Event` interface, which has only that method defined, and the second is to provide a unique event name to our application. For the domain events, they need to be unique within the bounded context in which they reside, but there is no harm in giving them a unique name that is also unique across an entire application.

Turning our attention to the **Ordering** module and the `CreateOrder` domain function, we will add a few lines just before `return` to add the event to the slice of aggregate domain events:

```
// … the rest of domain.CreateOrder()
order.AddEvent(&OrderCreated{
    Order: order,
})
return order, nil
}
```

Defining the events and updating the domain methods is easy enough, so we will go ahead and replicate how we just did it for `CreateOrder` and `OrderCreated`, and then do the same for `OrderCanceled`, `OrderReadied`, and `OrderCompleted`.

An interface is defined in the `application` package with methods for each of the preceding events:

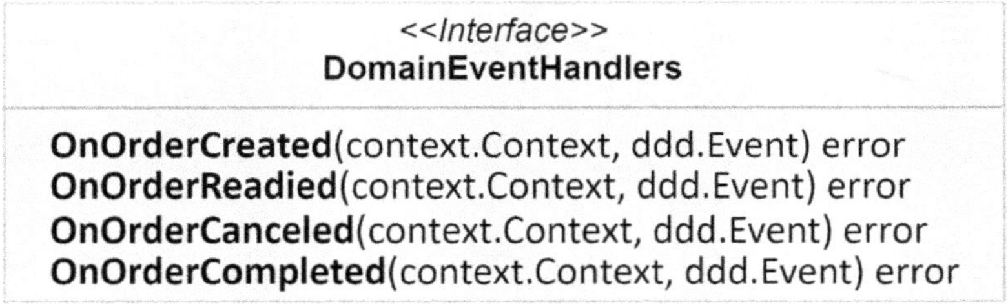

<<Interface>>
**DomainEventHandlers**

**OnOrderCreated**(context.Context, ddd.Event) error
**OnOrderReadied**(context.Context, ddd.Event) error
**OnOrderCanceled**(context.Context, ddd.Event) error
**OnOrderCompleted**(context.Context, ddd.Event) error

Figure 4.7 – The DomainEventHandlers interface

When this interface is implemented by `NotificationHandlers`, only three methods will be used. We can add an unused method to our implementation, but there is a slightly better alternative.

Consider for a moment a larger application with multiple handlers and a greater number of events, with an equal number of methods defined in `DomainEventHandlers`. Keeping each handler up to date would be tedious. We need a solution that will help us avoid creating empty and unused methods when new domain events have been added:

```
type ignoreUnimplementedDomainEvents struct{}
var _ DomainEventHandlers = (*ignoreUnimplementedDomainEvents)
(nil)
```

```
func (ignoreUnimplementedDomainEvents) OnOrderCreated( … )
error { … }
func (ignoreUnimplementedDomainEvents) OnOrderReadied( … )
error { … }
func (ignoreUnimplementedDomainEvents) OnOrderCanceled( … )
error { … }
func (ignoreUnimplementedDomainEvents) OnOrderCompleted( … )
error { … }
```

Because of the interface check, if DomainEventHandlers is changed when a new event is added, we will be alerted that ignoreUnimplementedDomainEvents is no longer in sync with those changes when we try to compile. We will avoid writing unused methods to keep up with the changes to DomainEventHandlers by including ignoreUnimplementedDomainEvents as an embedded field in our handlers:

```
type NotificationHandlers struct {
    notifications domain.NotificationRepository
    ignoreUnimplementedDomainEvents
}
```

The last new component to build is EventDispatcher:

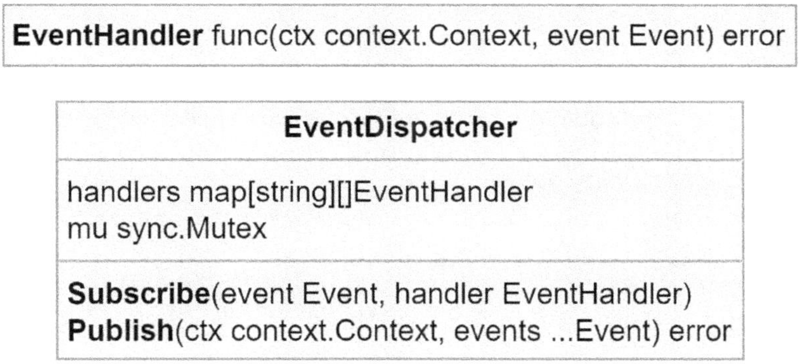

```
EventHandler func(ctx context.Context, event Event) error
```

```
EventDispatcher

handlers map[string][]EventHandler
mu sync.Mutex

Subscribe(event Event, handler EventHandler)
Publish(ctx context.Context, events ...Event) error
```

Figure 4.8 – EventDispatcher and EventHandler

EventDispatcher is nothing more than a simple implementation of the **Observer pattern** with its Subscribe and Publish methods:

```
func (h *EventDispatcher) Subscribe(
    event, handler EventHandler,
) {
```

```
        h.mu.Lock()
        defer h.mu.Unlock()
        h.handlers[event.EventName()] = append(
            h.handlers[event.EventName()],
            handler,
        )
    }
    func (h *EventDispatcher) Publish(
        ctx context.Context, events ...Event,
    ) error {
        for _, event := range events {
            for _, handler := range h.handlers[event.EventName()] {
                err := handler(ctx, event)
                if err != nil {
                    return err
                }
            }
        }
        return nil
    }
```

This new dispatcher and NotificationHandlers are brought together in a new RegisterNotificationHandlers function to create a driver adapter:

```
func RegisterNotificationHandlers(
    notificationHandlers application.DomainEventHandlers,
    domainSubscriber ddd.EventSubscriber,
) {
    domainSubscriber.Subscribe(
        domain.OrderCreated{},
        notificationHandlers.OnOrderCreated,
    )
    domainSubscriber.Subscribe(
        domain.OrderReadied{},
        notificationHandlers.OnOrderReadied,
    )
    domainSubscriber.Subscribe(
```

```
        domain.OrderCanceled{},
        notificationHandlers.OnOrderCanceled,
    )
}
```

The function accepts EventDispatcher with the EventSubscriber interface because we won't be needing the publication functionality here – at least not right now. When EventSubscriber is brought together with NotificationHandlers, subscriptions are made to the three events that the handlers are concerned with. With our ignoreUnimplementedDomainEvents solution, we can ignore making any subscriptions for events that we are not concerned with.

With all our components created and in place comes the time to add that plumbing I mentioned. To bring everything together, we head over to our composition root. Here is the modified composition root:

```
func (Module) Startup( … ) error {
    // setup Driven adapters
    domainDispatcher := ddd.NewEventDispatcher()
    …
    // setup application
    app = application.New(…, domainDispatcher)
    …
    // setup application handlers
    var notificationHandlers application.DomainEventHandlers
    notificationHandlers = application.
NewNotificationHandlers(notifications)
    …
    // setup Driver adapters
    …
    handlers.RegisterNotificationHandlers(
        notificationHandlers, domainDispatcher,
    )
    return nil
}
```

I'll break down what is happening in the preceding code:

- `EventDispatcher` is instantiated as `domainDispatcher` in the *Driven* section.
- We remove `notifications` from the parameter list for the application constructor and replace it with `domainDispatcher`. The application will not need to use the value any longer now that every use was moved into `NotificationHandlers`.
- We create an instance of `DomainEventHandlers` as `notificationHandlers`.
- The `notificationHandlers` instances are registered with `domainDispatcher` to create the subscriptions in the *Driver* section.

The final change that will be made is to each command handler deals with the creation, readiness, and cancelation of the order:

```
// ...
// publish domain events
if err = h.domainPublisher.Publish(
    ctx, order.GetEvents()...,
); err != nil {
    return err
}
```

Instead of using `notifications`, which are no longer available, we will publish the domain events generated within the `Order` aggregate. The preceding code snippet can be copied to each command handler without any modifications. The handlers are no longer responsible for or required to be aware of any potential side effects associated with the changes they end up making.

The additions made to the `ddd` package will make it easier to add domain event handling to other modules, and for the **Ordering** module, adding a second set of handlers to take care of the invoice side effects is also considerably easier.

## What about the modules not using DDD?

Not every module will need to use domain events. Forcing DDD onto a simple domain would be a counterproductive effort. When modules grow in complexity, they can be evaluated to determine whether refactoring and using DDD makes sense, but not before.

# Summary

In this chapter, you were shown around a monolith application and should now be familiar with the modules and the structure. You should also be able to run the application and use the UI to run experiments of your own. We also looked at how synchronous communication between components can work and the choices you might face when fetching data or sending commands. Then, in the last section, we implemented domain events in one of the modules. While this didn't change much or add any new asynchronous communication methods, it does set a foundation for us to build on to make not just our modules more reactive but the entire application.

In the next chapter, we will learn about **event sourcing** and implement it in the **Ordering** module. We will also cover event stores and CQRS.

# 5

# Tracking Changes with Event Sourcing

In the previous chapter, our MallBots application was updated to use events to communicate the side effects of changes to other parts of a module. These domain events are transient and disappear once the process ends. This chapter will build on the previous chapter's efforts by recording these events in the database to maintain a history of the modifications made to the aggregate.

In this chapter, we will be updating our domain events to support event sourcing, add an event sourcing package with new types, and create and use **Command and Query Responsibility Segregation** (CQRS) read models projected from our domain events. Finally, we will learn how to implement **aggregate snapshots**. Here is a quick rundown of the topics that we will be covering:

- What is event sourcing?
- Adding event sourcing to the monolith
- Using just enough CQRS
- Aggregate event stream lifetimes

By the end of this chapter, you will understand how to implement event sourcing along with CQRS read models in Go. You will also know how to implement aggregate snapshots and when to use them.

## Technical requirements

You will need to install or have installed the following software to run the application or try the examples:

- The Go programming language version 1.18+
- Docker

The code for this chapter can be found at `https://github.com/PacktPublishing/ Event-Driven-Architecture-in-Golang/tree/main/Chapter05`. Several modules in addition to the **Ordering** module used in the previous chapter have been updated to use domain events. In this chapter, we will be working in the **Store Management** module, namely `/stores`.

## What is event sourcing?

**Event sourcing** is a pattern of recording each change to an aggregate into an append-only stream. To reconstruct an aggregate's final state, we must read events sequentially and apply them to the aggregate. This contrasts with the direct updates made by a **create, read, update, and delete (CRUD)** system. In that system, the changed state of the record is stored in a database that overwrites the prior version of the same aggregate.

If we increase the price of a product, the following two tables show how that change might be recorded:

| Products | | | | | |
|---|---|---|---|---|---|
| ID | StoreID | Name | SKU | Price | ... |
| 1 | 1 | Diagrams for Dum... | BS-DD-123 | ***24.99*** | ... |
| 2 | 1 | 5001 Great Exam... | BS-GE-456 | 34.99 | ... |
| 3 | 2 | Wizard w/ Crystal | wizard | 28.99 | ... |
| ... | ... | ... | ... | ... | ... |

| Events | | | | | |
|---|---|---|---|---|---|
| ID | Type | Version | EventID | EventType | ... |
| 1 | Product | 1 | 101 | ProductAdded | ... |
| 1 | Product | 2 | 102 | ProductRebranded | ... |
| ***1*** | ***Product*** | ***3*** | ***103*** | ***ProductPriceIncreased*** | ... |
| ... | ... | ... | ... | ... | ... |

Figure 5.1 – A CRUD table (Products) and an event store table (Events)

When the price change has been saved to the **Products** table, only the price needs to change, leaving the rest of the row as is. We see in *Figure 5.1* that this is the case; however, we have lost both the previous price and the intent of the change.

The new price, as well as pertinent and valuable metadata, such as the purpose of the change, is saved when the change is recorded as an event in the **Events** table. The old price still exists in a prior event and can be retrieved if necessary.

Event sourcing implementations should use event stores that provide strong consistency guarantees and optimistic concurrency control. That is, when two or more modifications are made concurrently, only the first modification can add events to the stream. The rest of the modifications can be retried or would simply fail.

The event sourcing patterns can be used without any other event-driven patterns. It works very well with domain models (which use domain events) and event-driven architectures.

## Understanding the difference between event streaming and event sourcing

**Event streaming** is when events are used to communicate state changes with other bounded contexts of an application. Event sourcing is when events are used to keep a history of the changes in a single context and can be considered an implementation detail and not an architectural choice that has application-wide ramifications. These two uses of events are often thought to be the same and some speakers, books, and blogs conflate the two or use the terms interchangeably.

While event sourcing does use streams, as I mentioned at the start of *Chapter 1, Introduction to Event-Driven Architectures*, these streams are collections of events that are stored in a database that belongs to specific entities. Event streaming uses message brokers that have messages published to them and can be configured in a number of ways to then distribute those messages to consumers.

Additionally, the boundaries of the two are different. Event sourcing is implemented and contained within a single context boundary, whereas event streaming is typically used to integrate multiple context boundaries.

In terms of consistency models, an event streaming system is always going to be eventually consistent. An event-sourced system will have the same level of consistency as the database it is used with. With an ACID-compliant database, this would be strongly consistent. With non-relational databases, this is typically only eventually consistent. Even if event streaming is implemented within the same system as a strongly consistent event sourcing system, the former will not compromise the latter's level of consistency.

Event sourcing will require non-traditional thinking about your data. You will not be able to search for your data with complex queries, and you will need to build other ways to access your data besides simple identity lookups.

Most of all, event sourcing is also no silver bullet (a magical solution to a complicated problem). Event sourcing adds complexity to a system, and unlike a CRUD table, you will not be able to throw **object-relational mapping (ORM)** on top of it and call it a day.

### The importance of event sourcing in EDA

It is important to know that event streaming and event sourcing are different, but we should also know that they can work together and benefit from each other. They both benefit from the broader usage of events. The work that goes into breaking down the interactions on a domain or aggregate can be translated into events for event sourcing or as events that are going to be distributed to multiple systems in the application.

My intention is to explain event sourcing to you so that it is understood just as well as the event-driven patterns we will also cover. It is also useful to start with event sourcing since we will introduce a lot of concepts that will be reused throughout the book.

# Adding event sourcing to the monolith

In the previous chapter, we added a domain-driven design package for the monolith to use called ddd. We will need to make some updates to this package and add a new one for our event sourcing needs.

## Beyond basic events

The event code we used before was just what we needed. Those needs were to have them be easy to instantiate, be easy to reference as dependencies, and finally easy to work with in our handlers.

This is what we had before from *Chapter 4* in the *Refactoring side effects with domain events* section:

```
type EventHandler func(ctx context.Context, event Event) error

type Event interface {
    EventName() string
}
```

This old Event interface required that the **plain-old Go structs (POGSs)** that we are using implement the EventName() method to be seen by the application as an Event.

### Refactoring toward richer events

We have the following new needs:

- We need to know the details of which aggregate the event originates from
- Events need to be serialized and deserialized so they can be written and read back out of the database

With these new needs in mind, we need to revisit the events code in the ddd package and make some updates. The new interface for our event is shown in the following figure:

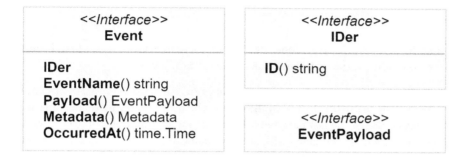

Figure 5.2 – The new Event and related interfaces

Looking at the new Event interface, some thoughts should spring to your mind. You might be thinking, for example, that there is no way you could add and manage all these methods to the events defined in the previous chapter (and you would be right).

What we used before as events will now be used as **event payloads**. The interface for EventPayload has no defined methods, which allows us to use it more easily. We might use an EventPayload of types bool or [] int if that is what fits best with our needs.

To create new events, we will use the following constructor:

```
type EventOption interface {
    configureEvent(*event)
}


func newEvent(
    name string, payload EventPayload,
    options ...EventOption,
) event {
    evt := event{
        Entity:     NewEntity(uuid.New().String(), name),
        payload:    payload,
        metadata:   make(Metadata),
        occurredAt: time.Now(),
    }
    for _, option := range options {
```

```
        option.configureEvent(&evt)
    }
    return evt
}
```

I will share what the new `event` struct looks like in a moment in *Figure 5.3*, but first, I want to take a quick moment to talk about the `options ...EventOption` parameter.

This variadic parameter will be used to modify the event prior to it being returned by the constructor as an `Event`. We will be using this to add in the details about the aggregate that creates the event. Variadic parameters must be the last parameter for a method or function and can be considered optional. The constructor could be called with an event name and a payload and nothing else. This technique is preferred over creating different variations of a constructor to handle combinations of parameters that might be used together.

> **Go 1.18 tip**
>
> We are beginning to use additions to the Go language that were added in the Go 1.18 release. In that release, the `any` alias was added for `interface{}`. Now, anywhere that we would have used a bare `interface{}`, we can replace it with the `any` alias. An example of `any` being used can be found in *Figure 5.3*. See `https://tip.golang.org/doc/go1.18` for more information on the changes that arrived in Go 1.18.

Now, back to the `event` struct. Here is what it looks like:

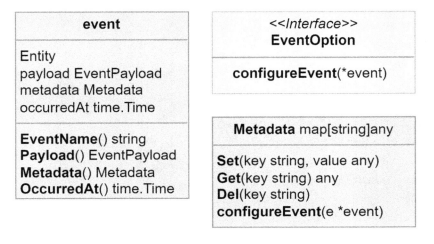

Figure 5.3 – The event and related types and interfaces

The event is a private struct that embeds Entity and contains only private fields. The use of EventOption will be the only way to modify these values outside of the ddd package.

> **Privacy in Go**
>
> Go does not have a private scope for types. The only kind of privacy scoping in Go is at the package level. To make something visible outside of the package, you *export* it by starting its name with an uppercase letter. Everything else that begins with a lowercase letter will be *unexported*. Types, constants, variables, fields, and functions can be made visible to other packages by being exported, but within a package, everything is always visible and accessible.

Events will now be capable of the following:

- Being created with additional metadata, such as the type and identity of the entity the event originated from
- Capturing the time when they occurred
- Performing equality checking based on their ID

That covers the updates made to the events themselves. While the change made may seem substantial, the work required to begin using them will be easy and mundane thanks to the foundation we created by adding domain events.

### Refactoring the aggregates to use the new rich events

Next up is the aggregate and it needs to be updated to use the new events constructor, among other small updates.

Here is our aggregate from the previous chapter:

```go
type Aggregate interface {
    Entity
    AddEvent(event Event)
    GetEvents() []Event
}
type AggregateBase struct {
    ID       string
    events []Event
}
```

By applying both an update and a bit of refactoring, we end up with this for `Entity` and `Aggregate`:

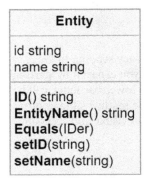
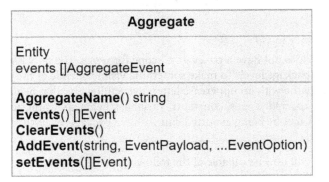

Figure 5.4 – The updated Entity and Aggregate and new AggregateEvent

In *Figure 5.4*, you might have noticed that the `AddEvent()` method signature looks much different from the previous version. This is the body of the updated `AddEvent()` method:

```
func (a *Aggregate) AddEvent(
    name string, payload EventPayload,
    options ...EventOption,
) {
    options = append(
        options,
        Metadata{
            AggregateNameKey: a.name,
            AggregateIDKey:   a.id,
        },
    )
    a.events = append(
        a.events,
```

```
        aggregateEvent{
            event: newEvent(name, payload, options...),
        },
    )
}
```

The AddEvent () method is not simply appending events to a slice any longer. We need to address the requirement to include the details about which aggregate the event originated from. To add the aggregate details, we use the Metadata type, which implements the EventOption interface. The method signature was also updated to accept the event name as the first parameter because the payloads no longer require an EventName () method, or methods for that matter.

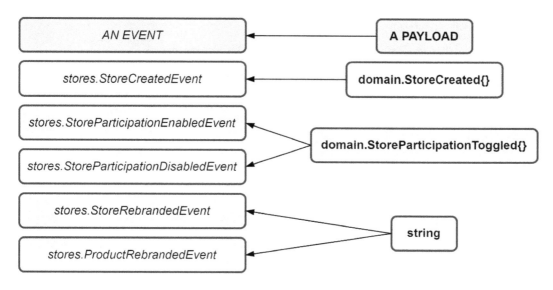

Figure 5.5 – Payloads may be used by multiple events

As *Figure 5.5* shows, the event name and the payload being separate means we can create a single payload definition, a struct, or a slice of integers, whatever we need, and use it with multiple events.

The updated `Aggregate` uses a new type of event, `AggregateEvent`, and the current `EventDispatcher` that we have only works with the `Event` type. Prior to Go 1.18, we had two choices:

- Create a new `AggregateEventDispatcher` to work with the new `AggregateEvent` type and keep type safety

- Use a single `EventDispatcher` and cast `Event` into `AggregateEvent` but lose type safety in the process

With Go 1.18 or greater, we can have both by updating `EventDispatcher` to accept a generic `Event` type:

Figure 5.6 – EventDispatcher and related types updated to use generic Event types

Now, let us see how we can update the monolith modules.

## Updating the monolith modules

I will be using the **Store Management** module to demonstrate the code updates that each module will receive. The rest I will leave for you or your compiler to explore on your own.

### Updating the aggregates

All of the locations in the monolith modules where new aggregates were instantiated will need to be modified to utilize the new `NewAggregate()` constructor. This modification is done in a few parts. We will create a constant to contain the aggregate name, create a constructor for the aggregate, and finally, replace each occurrence of aggregate instantiation.

For the `Store` aggregate, the following constructor is added:

```
const StoreAggregate = "stores.Store"

func NewStore(id string) *Store {
    return &Store{
```

```
        Aggregate: ddd.NewAggregate(id, StoreAggregate),
    }
}
```

Then, in `CreateStore()`, and in the PostgreSQL `StoreRepository` adapter, we update each place in the code where a new store is being created, for example, in `CreateStore()`:

```
store := NewStore(id)
store.Name = name
store.Location = location
```

Using the `NewStore()` constructor when creating `Store` aggregates in the future will ensure that the embedded `Aggregate` type is not left uninitialized.

## Updating the events

Starting with the store events, with matching changes for the products events, we move the return values from the `EventName()` methods for each store event to global constants:

```
const (
    StoreCreatedEvent              = "stores.StoreCreated"
    StoreParticipationEnabledEvent = "stores.
       StoreParticipationEnabled"
    StoreParticipationDisabledEvent = "stores.
StoreParticipationDisabled"
)
```

We add an `Event` suffix to each new constant, so we do not need to be bothered with renaming our payloads. When we are done creating the constants, the `EventName()` methods can all be removed. Next, we need to use these new constants as the first parameter in our calls to `AddEvent()`:

```
store.AddEvent(StoreCreatedEvent, &StoreCreated{
    Store: store,
})
```

After we make the changes to each `AddEvent()`, we have one final change to make before we can run the application again. There are no dramatic pauses in print, so if you want to take a guess, stop reading now before I reveal the change.

### Updating the event handlers

The handlers need to be updated to perform a type assertion on the value returned from the `Payload()` method on the event, and not on the event itself. A quick example from the notification handlers in the **Ordering** module is as follows:

```
func (h NotificationHandlers[T]) OnOrderCreated(
    ctx context.Context, event ddd.Event,
) error {
    orderCreated := event.Payload().(*domain.OrderCreated)
    return h.notifications.NotifyOrderCreated(
        ctx,
        orderCreated.Order.ID(),
        orderCreated.Order.CustomerID,
    )
}
```

After the updates to the handlers are complete, the monolith will compile again and run in the same manner. Here is an itemized list of the changes we made for updated events:

- Updated the `Event` interface and declared a new `EventPayload` interface
- Added an `event` struct and a new `Event` constructor
- Replaced the `Aggregate` interface with a struct and added a constructor for it
- Embedded an updated `Entity` into `Aggregate` and added a constructor for it as well
- Updated the `AddEvent()` method to track `Aggregate` information on the events
- Updated `EventDispatcher` to use generics to avoid losing type safety or creating many new versions
- Updated the modules to correctly build new `Aggregate` instances with new constructors
- Moved the event names into constants and used them in calls to `AddEvent()`
- Updated the handlers to perform type assertions on the event `Payload()`

The downside of updating the `ddd` package and making these changes is that the preceding changes affect any module that uses domain events and will need to be visited and updated. Next up is adding the code to support event sourcing our aggregates.

# Adding the event sourcing package

The updates made to enhance the events did not add any new code to support event sourcing. Because we do not want to turn every aggregate in the application into an event-sourced aggregate, the bits necessary to support event sourcing will go into a new `internal/es` package.

## Creating the event-sourced aggregate

To avoid having to build the aggregate from scratch, this new aggregate will contain an embedded `ddd.Aggregate` and provide a new constructor. Here is what we are starting with, the event-sourced `Aggregate` definition:

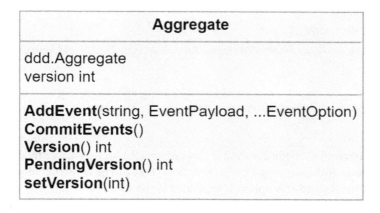

Figure 5.7 – The event-sourced Aggregate

This new `Aggregate` will also need a constructor:

```
func NewAggregate(id, name string) Aggregate {
    return Aggregate{
        Aggregate: ddd.NewAggregate(id, name),
        version:   0,
    }
}
```

The purpose of es.Aggregate struct is to layer on the versioning controls required to work with the event-sourced aggregates. It accomplishes this by embedding ddd.Aggregate. The AddEvent() method for the event-sourced Aggregate is defined as follows:

```go
func (a *Aggregate) AddEvent(
    name string,
    payload ddd.EventPayload,
    options ...ddd.EventOption,
) {
    options = append(
        options,
        ddd.Metadata{
            ddd.AggregateVersionKey: a.PendingVersion()+1,
        },
    )
    a.Aggregate.AddEvent(name, payload, options...)
}
```

We redefine the AddEvent() method so that it may decorate the options before they are passed into the same method from ddd.Aggregate. So that the events can be constructed with the correct version value, the ddd.Metadata option is appended to the slice of EventOption.

The constructors we added from the previous section for the aggregates, for example, Store and Product from the **Store Management** module, should switch from the ddd package to the es package so that the correct Aggregate constructor is called:

```go
func NewStore(id string) *Store {
    return &Store{
        Aggregate: es.NewAggregate(id, StoreAggregate),
    }
}
```

Then, after being updated to event-sourced aggregates, both `Store` and `Product` now have the lineage shown here:

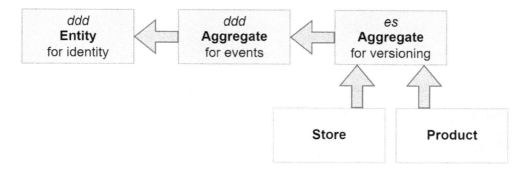

Figure 5.8 – The lineage of event-sourced aggregates

There is also an interface that needs to be implemented by each aggregate that is going to be event sourced:

```
type EventApplier interface {
    ApplyEvent(ddd.Event) error
}
```

This is how `ApplyEvent()` is implemented for `Product`:

```
func (p *Product) ApplyEvent(event ddd.Event) error {
    switch payload := event.Payload().(type) {
    case *ProductAdded:
        p.StoreID = payload.StoreID
        p.Name = payload.Name
        p.Description = payload.Description
        p.SKU = payload.SKU
        p.Price = payload.Price

    case *ProductRemoved:
        // noop

    default:
        return errors.ErrInternal.Msgf("%T received the event
        %s with unexpected payload %T", p, event.EventName(),
```

```
        payload)
    }
    return nil
}
```

I have chosen to keep the payloads aligned with the event names and can then use a `switch` statement that operates on the concrete types of the `EventPayload`. Another way would be using a `switch` statement and operating on the event names instead.

### Events are our source of truth

When an aggregate is event-sourced, the events are the source of truth. Put simply, what this means is that changes to the values within the aggregate should come from the events. Here is what `CreateStore()` should look like at this point:

```
func CreateStore(id, name, location string) (*Store, error) {
    if name == "" {
        return nil, ErrStoreNameIsBlank
    }
    if location == "" {
        return nil, ErrStoreLocationIsBlank
    }

    store := NewStore(id)
    store.Name = name
    store.Location = location

    store.AddEvent(StoreCreatedEvent, &StoreCreated{
        Store: store
    })
    return store, nil
}
```

The two parts that are highlighted, the assignments and the contents of the event, are what need to be changed:

- **Assignments**: We should not make any assignments directly. All assignments should be made by applying events to the aggregate. Domain functions and methods should emit events that contain information regarding a state change and should not directly modify the state on an aggregate.

- **Events**: We took some shortcuts with our domain events by passing around the aggregate in the events. Including the whole aggregate is perfectly fine for a domain event but we want to capture changes now and should be more specific about what goes into each event. This also means we cannot be as carefree about changing the contents of the events.

The assignments are simply removed, and the contents of the event are updated to only contain the `Name` and `Location` fields:

```
type StoreCreated struct {
    Name     string
    Location string
}
```

There will not be any need to add in the store ID because that is added by the `AddEvent()` method when it constructs the event with the information about the aggregate. The other store events, `StoreParticipationEnabled` and `StoreParticipationDisabled`, can be updated to be empty structs. Here is what the `ApplyEvent()` method looks like for the store:

```
func (s *Store) ApplyEvent(event ddd.Event) error {
    switch payload := event.Payload().(type) {
    case *StoreCreated:
        s.Name = payload.Name
        s.Location = payload.Location

    case *StoreParticipationEnabled:
        s.Participating = true

    case *StoreParticipationDisabled:
        s.Participating = false

    default:
        return errors.ErrInternal.Msgf("%T received the event
        %s with unexpected payload %T", s, event.EventName(),
        payload)
    }
    return nil
}
```

We will not be able to run the application now, and at this point, it should not even compile. There are some missing changes, such as accessing the now private entity `ID` value. Most of these issues will be within our repository adapters, which will be replaced shortly.

## Events need to be versioned

When what needs to be emitted from an aggregate needs to be changed, we cannot do it by changing the event. We could change the contents of an event when we were dealing with domain events because they are never persisted anywhere.

A new event (and maybe a new payload) needs to be created and that needs to be emitted from that point onward. The `ApplyEvent()` method will also need to keep handling the old event. When you use event sourcing, the application cannot forget the history of aggregates either.

### *Aggregate repositories and event stores*

Because we will be dealing with aggregates that regardless of their structure will be decomposed down to a stream of events, we can create and reuse a single repository and store.

Let us look at `AggregateRepository` and the interfaces involved in the following figure:

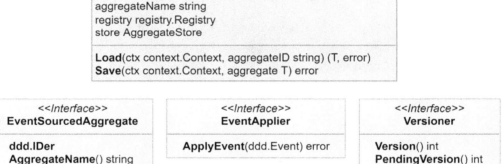

Figure 5.9 – AggregateRepository and related interfaces

`Load()` and `Save()` are the only methods we will use with event-sourced aggregates and their event streams. There are occasions when you would need to delete or alter events in the event store for reasons related to privacy or security concerns. This repository is not going to be capable of that and is not meant for that work. You would have some other specialized implementation you would use to gain access to the additional functions necessary. When working with event stores, securing them, and ensuring that they are in accordance with relevant legislation, such as the **General Data Protection Regulation (GDPR)**, can be challenging tasks.

For MallBots version 1.0, this repository is plenty sufficient.

It is important to understand what each method in `AggregateRepository` does.

The `Load()` method will do the following:

- Create a new concrete instance of the aggregate using the `registry.Build()` method
- Pass the new instance into the `store.Load()` method so it can receive deserialized data
- Return the aggregate if everything was successful

The `Save()` method will do the following:

- Apply any new events the aggregate has created onto itself
- Pass the updated aggregate into the `store.Save()` method so that it can be serialized into the database
- Update the aggregate version and clear the recently applied events using the `aggregate.CommitEvents()` method
- Return `nil` if everything was successful

### The data types registry

Looking into the `Load` method, we see the registry field in action:

```
func (r AggregateRepository[T]) Load(
    ctx context.Context, aggregateID, aggregateName string,
) (agg T, err error) {
    var v any
    v, err = r.registry.Build(
        r.aggregateName,
        ddd.SetID(aggregateID),
        ddd.SetName(r.aggregateName),
    )
    if err != nil { return agg, err }

    var ok bool
    if agg, ok = v.(T); !ok {
        return agg, fmt.Errorf("%T is not the expected type
        %T", v, agg)
    }
    if err = r.store.Load(ctx, agg); err != nil {
```

```
        return agg, err
    }
    return agg, nil
}
```

This code uses a type of registry to build a new instance of an aggregate and accepts two optional `BuildOption` parameters to set the `ID` and `Name` values of the aggregate that it builds.

| <<*Interface*>> **Registry** |
| --- |
| **Serialize**(key string, v interface{}) ([]byte, error)<br>**Build**(key string, options ...BuildOption) (interface{}, error)<br>**Deserialize**(key string, data []byte, options ...BuildOption) (interface{}, error) |

Figure 5.10 – The registry interface

This registry is useful for any event-sourced aggregate we need to deal with that happens to use the same event store as some others. To make that possible, we need a way to retrieve not just an interface but also the actual concrete type so that when `Load` returns, the caller is able to receive the correct type safely. The registry is very much like a prototype registry except it does not return clones of the original object.

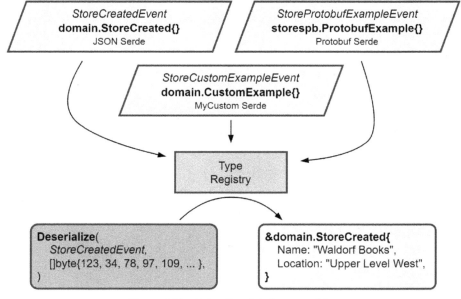

Figure 5.11 – Using the data types registry

The first step in using the registry is to register the types – zero values work best – that we want to retrieve later. The registry is very helpful in dealing with the serialization and deserialization of the data when we interact with a database or message broker. Each object that is registered is registered along with a serializer/deserializer, or serde for short. Different serdes can be used for different groups of objects. Later, when you interact with the registry to serialize an object, the registered serde for that type will perform the right kind of serialization. The same goes for `Build()` and `Deserialize()`; you will not need to know what kind of deserialization is at work to get your data turned into the right types again.

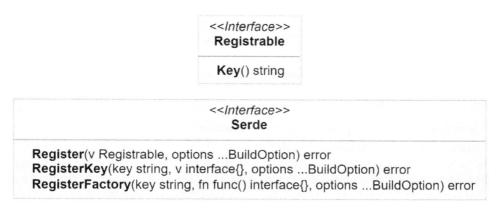

Figure 5.12 – The Registrable and Serde interfaces

Code that uses the instances created by the registry from serialized data will not need any modifications to work with the returned instances. The results from the registry are indistinguishable from instances that have never been serialized into a byte slice. This is the reason why the registry is used. The alternatives are managing mappers for each type or giving into the unknown and using the worst data structure in Go to work with: the dreaded `map[string]interface{}`.

When the registry is expected to work with more complex results that contain private fields, we need to reach for a `BuildOption` that has been defined in the same package as the type of the result we expect. That was the case in the example listing for the `Load` method. The private fields in the aggregate type were being set with `ddd.SetID()` and `ddd.SetName()`.

## *Implementing the event store*

`AggregateRepository` sits on top of an `AggregateStore`, which exists only as an interface.

Figure 5.13 – The AggregateStore interface

`AggregateStore` would be what finally makes the connection with the infrastructure on the other side. We will use this interface to create an event store that works with PostgreSQL.

> **Why have both AggregateRepository and AggregateStore?**
>
> It is reasonable to wonder at this point why both exist when it appears they both do the same thing. The repository has a few housekeeping actions, such as building the aggregate by left folding over the events or marking new events committed after a successful save, that must be taken care of for event sourcing to work, and the stores need to be implemented for specific infrastructure, such as an adapter. Instead of expecting each store implementation to do the tasks the repository does, the separate concerns are split into two parts.

### The events table DDL

The SQL is not complicated and can be easily modified to work with just about any relational database:

```
CREATE TABLE events (
  stream_id       text        NOT NULL,
  stream_name     text        NOT NULL,
  stream_version  int         NOT NULL,
  event_id        text        NOT NULL,
  event_name      text        NOT NULL,
  event_data      bytea       NOT NULL,
  occurred_at     timestamptz NOT NULL DEFAULT NOW(),
  PRIMARY KEY (stream_id, stream_name, stream_version)
);
```

This table should be added to the stores schema in the database. The CRUD tables, `stores` and `products`, should remain. We will have use for them in the next section.

In the `events` table, we use a compound primary key for optimistic concurrency control. Should two events come in at the same time for the same stream (`id` and `name`) and version, the second would encounter a conflict. As mentioned earlier, the application could try to redo the command and try saving again or give up and return an error to the user.

## Updating the monolith modules

We can start to plan out the changes we need to make to the composition root of our modules with event-sourced aggregates:

- We need an instance of the registry
- We need an instance of the event store
- We need new `Store` and `Product` repositories

### Adding the registry

The aggregate store and new `Store` and `Product` aggregate repositories will all need a registry. To be useful, that registry must contain the aggregates and events that we will be using. We can use a JSON `serde` because none of the domain types are complicated:

```
reg := registry.New()
err := registrations(reg)

func registrations(reg registry.Registry) error {
    serde := serdes.NewJsonSerde(reg)

    serde.Register(domain.Store{})
    serde.Register(domain.StoreCreated{})
}
```

### Adding the event store

There is nothing complicated about creating the event store instance:

```
eventStore := pg.NewEventStore("events", db, reg)
```

### Replacing the aggregate repositories

We will need to update the repository interfaces for the Store and Product aggregates. The following is the updated interface for StoreRepository and the new repository for Product will be very similar:

Figure 5.14 – The event-sourced StoreRepository interface

There will not be any need to write any new implementations for the two new interfaces. AggregateRepository uses generics, and we can again have type safety and save a little on typing. To create a new instance of StoreRepository, we replace the previous stores instantiation with the following line in the composition root:

```
stores := es.NewAggregateRepository[*domain.Store](
    domain.StoreAggregate,
    reg,
    eventStore,
)
```

When we make this change, our repository interface will not be sufficient to handle queries that return lists or involve filtering. Event sourcing is not going to be useful for, and, in most cases, will be impossible to use with, these kinds of queries.

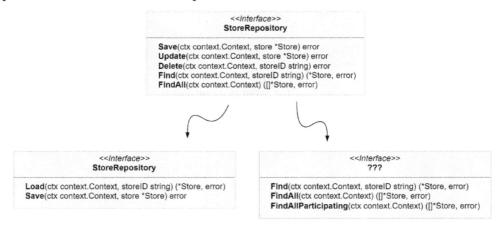

Figure 5.15 – Breaking up the old StoreRepository interface

If we update the `StoreRepository` interface, as shown in *Figure 5.15*, there will still be several methods we need to implement with an unknown interface and data source that the event store is incapable of doing. This limitation is the reason why CQRS is introduced in most systems that use event sourcing. CQRS can be implemented without event sourcing, but it is difficult to implement event sourcing without CQRS. We will need to create some read models for the queries that the event store will be unable to handle. It will mean some more work, but we are prepared. The foundation we made in the previous chapter with domain events is going to make that work much easier.

# Using just enough CQRS

The **Store Management** module has a number of existing queries in the application. Some we may be able to serve from the event store, such as `GetProduct()` and `GetStore()`, but the others, such as `GetParticipatingStores()` or `GetCatalog()`, would require scanning the entire table to rebuild every stream, and then we would filter a percentage out.

When we created the events table in the previous section, we left the existing tables alone. This was a tiny bit of cheating on my end. Although I knew the tables would be used again for our read models, it might not always be practical to reuse old tables. In most cases, the tables that support your read models should be specifically designed to fulfill requests as efficiently as possible. The tables that are left over after a refactoring might not be suitable for that task.

We could also use entirely new tables, use a new database, and even do more beyond using different read models. Right now, our only need is to get the queries working again and the discarded `stores` and `products` tables will do the job and already exist. There is also some discarded code we can repurpose to make the job of creating our read models go quicker.

## A group of stores is called a mall

A new interface, `MallRepository`, needs to be created to house all the queries that `StoreRepository` will be unable to handle. To create the read model, we will need to project the domain events into it with an event handler.

This is the `MallRepository` interface that will require a PostgreSQL implementation:

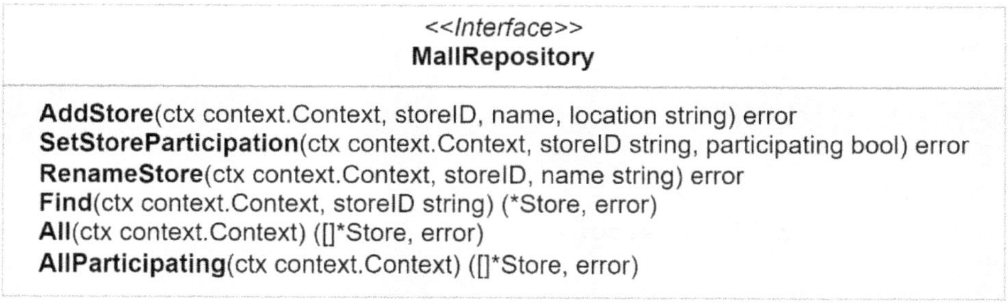

| <<Interface>> |
| --- |
| **MallRepository** |
| **AddStore**(ctx context.Context, storeID, name, location string) error <br> **SetStoreParticipation**(ctx context.Context, storeID string, participating bool) error <br> **RenameStore**(ctx context.Context, storeID, name string) error <br> **Find**(ctx context.Context, storeID string) (*Store, error) <br> **All**(ctx context.Context) ([]*Store, error) <br> **AllParticipating**(ctx context.Context) ([]*Store, error) |

Figure 5.16 – The MallRepository interface

This repository is concerned with projecting events related to stores into the read model and performing those queries that cannot be handled by the event store. Much of the code from the previous iteration of the `StoreRepository` PostgreSQL implementation can be migrated to the PostgreSQL implementation of `MallRepository`.

The new repository needs to be created in the composition root so that it can be passed into the application and used in place of `StoreRepository` and `ParticipatingStoreRepository` in the queries:

```
mall := postgres.NewMallRepository("stores.stores", mono.DB())
// ...
application.New(stores, products, domainDispatcher, mall)
```

The new `MallRepository` is also used in the application in place of `StoreRepository`:

```
// ...
appQueries: appQueries{
    GetStoreHandler:
        queries.NewGetStoreHandler(mall),
    GetStoresHandler:
        queries.NewGetStoresHandler(mall),
    GetParticipatingStoresHandler:
        queries.NewGetParticipatingStoresHandler(mall),
    GetCatalogHandler:
        queries.NewGetCatalogHandler(products),
    GetProductHandler:
        queries.NewGetProductHandler(products),
},
// ...
```

The query handlers will all also need to be updated so that they accept `MallRepository` instead of either `StoreRepository` or `ParticipatingStoreRepository` and also update any method calls to the correct ones; for example, this is the `GetStores` handler:

```
type GetStores struct{}

type GetStoresHandler struct {
    mall domain.MallRepository
}
```

```go
func NewGetStoresHandler(mall domain.MallRepository)
GetStoresHandler {
    return GetStoresHandler{mall: mall}
}


func (h GetStoresHandler) GetStores(
    ctx context.Context, _ GetStores,
) ([]*domain.Store, error) {
    return h.mall.All(ctx)
}
```

The last thing to do for this new read model and its projections is to add the event handlers. Before we do, I need to share a small bit of behind-the-scenes refactoring that needs to happen regarding the event dispatcher and handlers.

### Refactoring out the extra steps to handle events

In the last chapter, I shared a technique that is used to avoid having to implement event handlers, the `ignoreUnimplementedDomainEvents` embed. It helped, but as I was going over the code to add logging, it became obvious I was still having to deal with implementing a lot of methods. I was also adding new events to test the event sourcing in the **Store Management** module and that meant more methods to implement. The solution was to use a common interface to handle all events, not just as a `func` type, but as a true interface in `internal/ddd/event_dispatcher.go`.

Figure 5.17 – The new EventHandler and EventHandlerFunc types

The old `EventHandler` from before still exists but as `EventHandlerFunc`. Now, either a value that implements `EventHandler` can be passed into `EventDispatcher.Subscribe()` or a func that has the correct signature can be passed in, as in this example:

```go
func myHandler(ctx context.Context, ddd.Event) error {
    // ...
    return nil
}
dispatcher.Subscribe(
    MyEventName,
    ddd.EventHandlerFunc(myHandlerFn),
)
```

This may seem familiar; the technique is rather common, and you might have even encountered it from the `http` package in the standard library where it is used to allow the creation of router handlers with implementations of `http.Handler` or by wrapping any `func(http.ResponseWriter, *http.Request)` with `http.HandlerFunc`.

The `EventHandler` interface update makes the logging for the event handlers much less of a chore with only one method that needs to exist to log all accesses:

```go
type EventHandlers struct {
    ddd.EventHandler
    label  string
    logger zerolog.Logger
}

var _ ddd.EventHandler = (*EventHandlers)(nil)

func (h EventHandlers) HandleEvent(
    ctx context.Context, event ddd.Event,
) (err error) {
    h.logger.Info().Msgf(
        "--> Stores.%s.On(%s)",
        h.label,
        event.EventName(),
    )
    defer func() {
        h.logger.Info().Err(err).Msgf(
```

```
            "<-- Stores.%s.On(%s)",
            h.label,
            event.EventName(),
        )
    }()
    return h.EventHandler.HandleEvent(ctx, event)
}
```

The application.DomainEventHandlers in each module was also removed. The handlers provided protection from panics when we encountered an event without a designated handler. In the future, unhandled events will not result in any panics, and we do not require this protection.

### *Adding the mall event handlers*

After the EventHandler refactoring, the handlers have just one method to implement and just like ApplyEvent(), we are free to choose how to implement it. Opposite to the way I am doing it in the aggregates, because event payloads can be shared by different events, I find it is easiest to use a switch that operates on the event names in these handlers:

```
type MallHandlers struct {
    mall domain.MallRepository
}


var _ ddd.EventHandler = (*MallHandlers)(nil)


func (h MallHandlers) HandleEvent(
    ctx context.Context, event ddd.Event,
) error {
    switch event.EventName() {
    case domain.StoreCreatedEvent:
        return h.onStoreCreated(ctx, event)
    case domain.StoreParticipationEnabledEvent:
        return h.onStoreParticipationEnabled(ctx, event)
    case domain.StoreParticipationDisabledEvent:
        return h.onStoreParticipationDisabled(ctx, event)
    case domain.StoreRebrandedEvent:
        return h.onStoreRebranded(ctx, event)
    }
```

```
        return nil
    }
    // ...
```

The HandleEvent() method simply proxies the event into different methods based on the event name. I made the decision to call out the unexported methods in order to isolate the handling of each event from the others. By structuring it this way, I could more easily reuse the methods, but HandleEvent() could have any structure or style that gets the job done.

Post-refactoring, setting up the handler subscriptions is also a smidge easier. We need to create four subscriptions for the events in the preceding listing:

```
func RegisterMallHandlers(
    mallHandlers ddd.EventHandler,
    domainSubscriber ddd.EventSubscriber,
) {
    domainSubscriber.Subscribe(
        domain.StoreCreatedEvent, mallHandlers,
    )
    domainSubscriber.Subscribe(
        domain.StoreParticipationEnabledEvent,
        mallHandlers,
    )
    domainSubscriber.Subscribe(
        domain.StoreParticipationDisabledEvent,
        mallHandlers,
    )
    domainSubscriber.Subscribe(
        domain.StoreRebrandedEvent, mallHandlers,
    )
}
```

A last update to the composition root is to wire the preceding up like we have with handlers in the past and we are done adding the mall read model.

# A group of products is called a catalog

Adding the read model for the catalog will be handled in a very similar fashion to the mall read model. I will not be going over each part in the same detail but will instead provide `CatalogRepository`, the list of interesting events, some of which are new, and an itemized list of the changes.

This is the `CatalogRepository` interface:

| *<<Interface>>* **CatalogRepository** |
| --- |
| **AddProduct**(ctx context.Context, productID, storeID, name, description, sku string, price float64) error<br>**Rebrand**(ctx context.Context, productID, name, description string) error<br>**UpdatePrice**(ctx context.Context, productID string, price float64) error<br>**RemoveProduct**(ctx context.Context, productID string) error<br>**Find**(ctx context.Context, productID string) (*Product, error)<br>**GetCatalog**(ctx context.Context, storeID string) ([]*Product, error) |

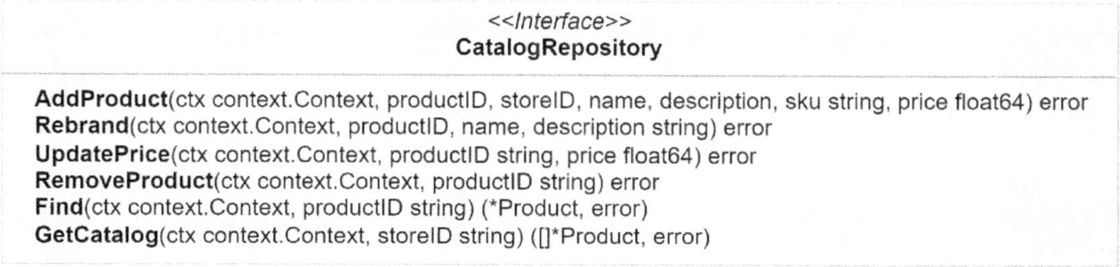

Figure 5.18 – The CatalogRepository interface

The events, existing and new ones, mean we have ended up with more modification-making methods than ones performing queries. We are focused on projecting the events into the read model and so the interface is designed to handle the data from each event we are interested in.

The interesting `Product` events that should be handled in the `CatalogHandlers` are as follows:

- `ProductAddedEvent`: An event that contains fields for each value set on a new product

- `ProductRebrandedEvent`: A new event that contains a new name and description for an existing product

- `ProductPriceIncreasedEvent`: A new event that contains a higher price for an existing product in the form of a price change delta

- `ProductPriceDecreasedEvent`: A new event that contains a new lower price for an existing product in the form of a price change delta

- `ProductRemovedEvent`: An empty event signaling the deletion of an existing product

The steps to connect the domain events with the catalog read model are as follows:

1. Implement `CatalogRepository` as a PostgreSQL adapter

2. Create an instance of the adapter in the composition root

3. Pass the instance into the application and replace the products event store with the catalog instance in each query handler

4.  Create `CatalogHandlers` in the application package with a dependency on `CatalogRepository`

5.  Create an instance of the handlers and repository in the composition root

6.  Pass them into the subscription handlers and subscribe for each event

## Taking note of the little things

There are still some little things that need addressing because of the CQRS read model changes. The application command handlers need a once-over to fix any calls to methods on the repositories that no longer exist. The `RemoveProduct` command handler, for example, needs to not call `Remove()` on the repository but it should instead be calling `Save()`, as weird as that may sound. This is because we will not be performing a `DELETE` operation in the database when we remove a product. Instead, a new `ProductRemovedEvent` will be appended to the event stream for the removed `Product` aggregate.

Another small issue is that the aggregate repository and the stores will not return an error if the stream does not exist. Most of the time, this will be alright; however, if what was returned was an empty aggregate and we are not expecting a fresh aggregate, then we need validations in place to keep events from being added and applied when they should not be.

## Connecting the domain events with the read model

Running the application now with all the changes in place for the **Store Management** module, we will see the following appear in the logs when we add a new product:

```
started mallbots application
web server started; listening at http://localhost:8080
rpc server started
INF --> Stores.GetStores
INF <-- Stores.GetStores
INF --> Stores.AddProduct
INF <-- Stores.AddProduct
```

Assuming everything is wired up correctly and the handler access logging is set to log all `HandleEvent()` calls, this may seem a little confusing. There should be some additional lines in there that show the `HandleEvent()` method on `CatalogHandlers` was accessed. What we should be seeing is this:

```
started mallbots application
web server started; listening at http://localhost:8080
rpc server started
INF --> Stores.GetStores
```

```
INF <-- Stores.GetStores
INF --> Stores.AddProduct
INF --> Stores.Catalog.On(stores.ProductAdded)
INF <-- Stores.Catalog.On(stores.ProductAdded)
INF <-- Stores.AddProduct
```

There is a simple explanation for why we do not see the event being handled. The reason we do not see the extra lines showing us that the CatalogHandlers implementation received the stores.ProductAdded event is that by the time the domain event publisher gets a hold of the product aggregate, the events have already been committed and cleared from it. Here are the important lines from the AddProduct command handler:

```
// ...
if err = h.products.Save(ctx, product); err != nil {
    return err
}
if err = h.domainPublisher.Publish(
    ctx, product.GetEvents()...,
); err != nil {
    return err
}
// ...
```

Recall that the third step for the Save() method on AggregateRepository is *update the aggregate version and clear the events using the aggregate.CommitEvents() method.*

Moving the Publish() call before the Save() call would seem to be the answer if the issue is that the events are cleared within Save(). This can work as long as the following apply:

- Everyone remembers that Publish() must always precede Save()

- There are never any issues in applying the events causing Save() to fail

Another answer would be to make the publishing action part of the saving action. No one would need to remember which action needed to be first and the errors from the Save() method can be automatically handled. Another bonus to having it as part of the saving action is the command handlers would no longer need the publisher (or be concerned with the publishing action).

We could modify AggregateRepository to depend on the EventPublisher interface and have it take care of the publishing before it commits the events. We could then have Publish() before or after Save(). This would be coupling the repository and publisher together. If we wanted to not do any publishing and did not want to provide the publisher, we would need to pass in nil to the constructor and check for a nil publisher before calling methods on it.

We could use a variadic parameter to pass in the publisher and other options if we had any. This would improve the situation with passing in a nil parameter, but we would still need to perform a check on the publisher before using it.

Either option would be straightforward enough to implement quickly but they both suffer from having to make modifications to AggregateRepository that create a dependency on a publisher.

### Doing more with middleware

The better solution I see for this situation is to use the Chain of Responsibility pattern (https://en.wikipedia.org/wiki/Chain-of-responsibility_pattern) to chain AggregateStore calls together. You may know this pattern by its more common term, middleware. With this solution, there will be no modifications made to AggregateRepository; likewise, no changes are made to the AggregateStore interface or the EventStore implementation.

This is actually not very different than what we are doing right now with logging. LogApplicationAccess to wrap each application call to add some simple logging.

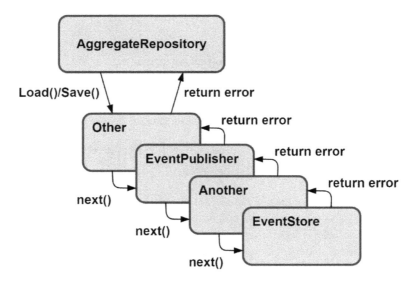

Figure 5.19 – Using middleware with AggregateRepository

Each middleware in the preceding example, `Other`, `EventPublisher`, and `Another`, returns some handler that implements the same `AggregateStore` interface that `EventStore` does.

To build the chain, each middleware constructor returns an `AggregateStoreMiddleware` function that has the following signature:

```
func(store AggregateStore) AggregateStore
```

To build our chain, we need a function that takes an `AggregateStore` interface and then accepts a variable number of `AggregateStoreMiddleware` functions. What it will do is execute a loop over each middleware, in reverse order, passing in the result from the previous loop:

```
func AggregateStoreWithMiddleware(
    store AggregateStore, mws ...AggregateStoreMiddleware,
) AggregateStore {
    s := store
    for i := len(mws) - 1; i >= 0; i-- {
        s = mws[i](s)
    }
    return s
}
```

If we were to call the preceding function with `store, A, B, C`, the result we could get back would be `A(B(C(store)))`. Most of the time, the order we add middleware is not much of a concern because the middleware are not able to work together, or it is strongly suggested they do not, but there are some kinds of middleware that we do want to be closer to either the end of the chain or the beginning. An example might be middleware that recovers from panics in the code. We would want to have that middleware at the very start of the chain so that any panic generated anywhere in the chain is caught and recovered from.

The only middleware we have right now is the one for `EventPublisher`. It will need to publish the events in the `Save()` call either before or after it makes its call to the next store in the chain. We will not need to take any action on a call to `Load()`, so it makes sense to use an embedded `AggregateStore` so we can avoid having to write a proxy method we will not be doing anything with:

```
type EventPublisher struct {
    AggregateStore
    publisher ddd.EventPublisher
}


func NewEventPublisher(publisher ddd.EventPublisher)
  AggregateStoreMiddleware {
```

```
        eventPublisher := EventPublisher{
            publisher: publisher,
        }
        return func(store AggregateStore) AggregateStore {
            eventPublisher.AggregateStore = store
            return eventPublisher
        }
    }
}

func (p EventPublisher) Save(
    ctx context.Context, aggregate EventSourcedAggregate,
) error {
    if err := p.AggregateStore.Save(ctx, aggregate); err != nil
{
        return err
    }
    return p.publisher.Publish(ctx, aggregate.Events()...)
}
```

The middleware and the constructor for it are shown in the previous code block. Highlighted is the actual middleware function that will be used by the chain builder `AggregateStoreWithMiddleware()` function.

To use the middleware, we need to update the composition root for the **Store Management** module by surrounding the creation of the store with the chain builder:

```
aggregateStore := es.AggregateStoreWithMiddleware(
    pg.NewEventStore("events", db, reg),
    es.NewEventPublisher(domainDispatcher),
)
```

Domain events will always be published after the events have been successfully persisted into the database.

---

**This is still not event streaming**

We only have an `AggregateStore` middleware that helps us with the issue of publishing domain events when we make a change to an event-sourced aggregate. Everything is still very much contained within our bounded context and is still synchronous. In later chapters, when we add asynchronous integrations; the aggregate repositories or stores will not be involved.

Now that `EventDispatcher` has been added using middleware to the store used by the aggregate repositories, the application command handlers no longer need to depend on it. Any places in the application and commands packages that reference `ddd.EventPublisher` should be updated to remove the reference. Under normal circumstances, they will not be doing anything because the events will be cleared, but when things go wrong, they may still publish events we would not want to be published.

### Recapping the CQRS changes

We added, updated, or implemented the following things:

- Read models called `MallRepository` and `CatalogRepository` were created
- The application was updated to use the read models in the query handlers
- The event handler signature was refactored to reduce boilerplate
- We added support for adding middleware to `AggregateStore`
- We used middleware to publish domain events

The decision to use event sourcing on a module that has existing queries forced our hand and we had to implement read models. We took a shortcut and reused the tables we had just discarded and that ended up saving a lot of effort.

## Aggregate event stream lifetimes

In an event-sourced system, there are two kinds of aggregates:

- Short-lived aggregates, which will not see many events in their short lifetime
- Long-lived aggregates, which will see many events over their very long lifetime

Examples of a short-lived aggregate would be `Order` from the **Ordering** module and `Basket` from the **Shopping Baskets** module. Both exist for a short amount of time, and we do not expect them to see many events. Examples of long-lived aggregates are `Store` from the **Store Management** module and `Customer` from the **Customers** module. These entities will be around for a long time and can end up seeing many events.

The performance of short-lived aggregates, and streams with few events in general, is not going to be a problem. The small number of events can be read and processed quickly. Larger streams would take longer to read and process; the larger it is, the longer it would take.

## Taking periodic snapshots of the event stream

When we know that we will be dealing with a larger stream, we can use snapshots to improve performance by reducing the number of events we will load and process. In *Figure 5.20*, the state of the stream is saved along with the aggregate version.

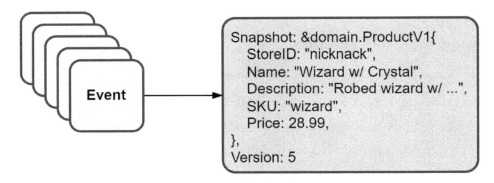

Figure 5.20 – Capturing the current state of the event stream as a snapshot

A snapshot is a serialization of the aggregate and the version of the aggregate it came from. When we create the serialization, we do not want to create it from the aggregate because that would limit the flexibility to change the structure of the aggregate in the future. Instead, we should use versioned representations, that is, `ProductV1`, which is then serialized and saved.

An aggregate that is going to use snapshots will need to implement the `ApplySnapshot()` and `ToSnapshot()` methods from the `Snapshotter` interface.

Figure 5.21 – The Snapshotter and Snapshot interfaces

Adding snapshots to an aggregate does not require making any other changes to the aggregate or to the constructor that builds it outside of the two new methods to satisfy the `Snapshotter` interface.

A snapshot should have everything to recreate the aggregate exactly as it was. In most cases, this means the same structure is duplicated as a `Snapshot` struct. The aggregate can continue to evolve, and older snapshots can still be loaded in `ApplySnapshot()` by using a switch that operates on either the type or name. For example, this is the switch statement used for the `Product` aggregate:

```
switch ss := snapshot.(type) {
case *ProductV1:
    p.StoreID = ss.StoreID
    p.Name = ss.Name
    p.Description = ss.Description
    p.SKU = ss.SKU
    p.Price = ss.Price

default:
    return errors.ErrInternal.Msgf("%T received the unexpected
snapshot %T", p, snapshot)
}
```

If `Product` were modified tomorrow and a new field, `Weight int`, was added, then a new `Snapshot` should be created and called `ProductV2`. It should contain all the fields from `ProductV1` and a new one for `Weight`. `ToSnapshot()` would be updated to return a `ProductV2` snapshot going forward, and `ApplySnapshot()` should be updated to handle the new snapshot as well. The code to handle the older snapshot version(s) may or may not need to be modified. In this example case, there would be no modification needed. If there is never any event that modifies the `Weight` value of a `Product`, then the zero value, in this case, literally zero, will be used as the value for `Weight`. When `Weight` is finally given a non-zero value, it should not be assumed that the old snapshot will be replaced as well. An older snapshot may continue to exist in the database if the snapshot strategy did not signal that a new one should be taken, causing it to be replaced.

### Strategies for snapshot frequencies

How often you take a snapshot is subject to the strategy you use. Some strategy examples are as follows:

- *Every N events* strategies create new snapshots when the length of loaded or saved events has reached some limit, such as every 50 events

- *Every period* strategies create new snapshots every new period, such as once a day or every hour

- *Every pivotal event* strategies create a snapshot when a specific event is appended to the stream, such as when a store is rebranded

Your choice of strategy should be guided by the business needs of the aggregate and domain.

> **Hardcoded strategy used in the book's code**
> The code shared for this book uses a strategy that will create a snapshot every three events. Every three events is not a good strategy outside of demonstration purposes.

## Using snapshots

There is not going to be a special interface for snapshots; a PostgreSQL SnapshotStore that satisfies the AggregateStore interface is used. To make easy work of both applying and taking snapshots, we turn to AggregateStoreMiddleware again.

### The snapshots table DDL

Another simple CREATE TABLE statement that could work with other relational databases is as follows:

```
CREATE TABLE baskets.snapshots (
    stream_id        text        NOT NULL,
    stream_name      text        NOT NULL,
    stream_version   int         NOT NULL,
    snapshot_name    text        NOT NULL,
    snapshot_data    bytea       NOT NULL,
    updated_at       timestamptz NOT NULL DEFAULT NOW(),
    PRIMARY KEY (stream_id, stream_name)
);
```

The primary key of the snapshots table is not like the one in events. New versions of a snapshot will overwrite older ones using an UPDATE statement. It does not have to work this way and the primary key could be changed to keep a history of snapshots if desired.

### Plugging into the aggregate store middleware

SnapshotStore could be coded to stand alone but the implementation that is being used in this application is coded up to work as AggregateStoreMiddleware. Here is the store middleware statement from the **Store Management** module with the new snapshot middleware added:

```
aggregateStore := es.AggregateStoreWithMiddleware(
    pg.NewEventStore("stores.events", mono.DB(), reg),
    es.NewEventPublisher(domainDispatcher),
    pg.NewSnapshotStore("stores.snapshots", mono.DB(), reg),
)
```

## Loading aggregates from snapshots

When `AggregateRepository` is executing `Load()` for an aggregate, the middleware will check for a snapshot, and if one is found, it will apply it using the `ApplySnapshot()` method. The modified aggregate is then passed to the next `Load()` handler.

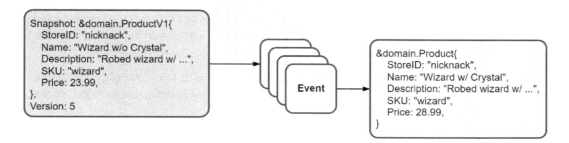

Figure 5.22 – Using a snapshot to load fewer events

The event store will then load the events from the aggregate's current version, which skips all events that have a version equal to or lower than what the snapshot has for its version.

If you rush over to test the application and try a `GetProduct()` query, you have forgotten that those use a read model. To test that aggregate snapshot functionality is working, you will need to look into the `snapshots` table after making some changes to a `Product` or `Store` aggregate. If you see rows appearing and repeated modifications continue, then everything is working as intended.

### *Snapshots are not without their downsides*

A snapshot is a type of cache, and like other caches, it has various downsides, such as the duplication and invalidation of data. Snapshots are a performance optimization that should only be used when absolutely necessary. Otherwise, you would be making a premature optimization. Your snapshots would also be subject to the same security, legal, and privacy considerations that you might have to make for the events.

# Summary

We covered a lot about event sourcing and went into a lot of the interfaces, structs, and types used to create an event sourcing implemention in Go. We started off by making a pretty big change to the simple events model used in the last chapter. This was followed by updates to the aggregate model and an entirely new package.

We also learned about a type of registry for recallable data structures and how it is implemented and used. Refactoring for event handlers was introduced, which shaved a good number of lines from the repository, which is always a good thing.

Introducing CQRS and implementing read models could not be avoided, but working through it and implementing it revealed it to not be such a confusing or complicated pattern, thanks in part to the work from the previous chapter, of course.

We closed out the chapter by implementing snapshots in the application and covered why and when you would use them in your own applications.

I did mention twice, and this makes the third time, that what we were doing with event sourcing is not considered by some to be event-driven architecture because event sourcing is not a tool for integrating domains. Regardless, the pattern involved events, and it allowed me to introduce richer event models before also introducing messages, brokers, and asynchronous communication.

In the next chapter, *Chapter 6, Asynchronous Connections*, we will learn about messages, brokers, and, of course, finally adding asynchronous communication to the application.

# 6
# Asynchronous Connections

The events we have worked with so far in this book have been synchronously handled. The domain events in *Chapter 4*, *Event Foundations*, were used to move the side effects of domain-model changes into *external handlers*.

External handlers were called after the change was made successfully and within the same process. In *Chapter 5*, *Tracking Changes with Event Sourcing*, we used events to record each change made to our domain aggregates. When we want to use an aggregate, we read all of the events in sequence to rebuild the current state of the aggregate. With both kinds of events, our system is immediately or strongly consistent because events are always created or read within a single process.

We will be covering the following topics in this chapter:

- Asynchronous integration with messages

- Implementing messaging with **NATS JetStream**

- Making the **Store Management** module asynchronous

The events we will be working with in this chapter and for the remainder of the book will be asynchronous. The umbrella term for these events is integration events. Both **notification** and **event-carried state transfer** events are types of integration events.

## Technical requirements

In this chapter, we will be adding asynchronous messaging to some modules using **Neural Autonomic Transport System** (**NATS**) JetStream. You will need to install the following software to run the application and to try the examples specified in the chapter:

- The Go programming language version 1.18+

- Docker

The source code for the version of the application used in this chapter can be found at `https://github.com/PacktPublishing/Event-Driven-Architecture-in-Golang/tree/main/Chapter06`.

## Asynchronous integration with messages

So far in this book, we have only talked about events, so what exactly is a **message**? An event is a message, but a message is not always an event. A message is a container with a payload, which can also be an event and can have some additional information in the form of key-value pairs.

A message may be used to communicate an event, but it may also be used to communicate an instruction or information to another component.

The kinds of payloads we will be using in this book include the following:

- **Integration event**: A state change that is communicated outside of its bounded context
- **Command**: A request to perform work
- **Query**: A request for some information
- **Reply**: An informational response to either a command or query

The first kind of message we will be introduced to and will implement is an **integration event**. The term *integration event* comes from how it is used to integrate domains and bounded contexts. This is how an integration event compares with the domain and event-sourced events we have already worked with:

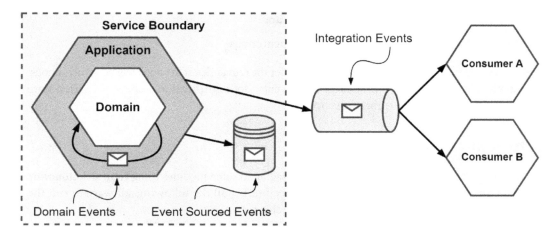

Figure 6.1 – Event types and their scopes

An application uses different kinds of events to accomplish a variety of activities:

- **Domain events**: Exist for the shortest time, never leave the application, do not require versioning, and are typically handled synchronously. These events are used to inform other application components about changes made to or by an aggregate.

- **Event-sourced events**: Exist for the longest time, never leave the service boundary, require versioning, and are handled synchronously. These events keep a record of every change in state that is made to an aggregate.

- **Integration events**: Exist for an unknown amount of time, are used by an unknown number of consumers, require versioning, and are typically handled asynchronously. These events are used to supply other components of the system with information regarding significant decisions or changes.

Both notification and event-carried state transfer events are integration events, as mentioned before.

## Integration with notification events

A **notification event** is going to be the smallest event you can send. You might send a notification because the volume of the event is very high, or you might send one because the size of the data related to the change is too large.

Some examples of when to use a notification are presented here:

- New media has been uploaded or has become available. Serializing the file content into an event is not likely to be practical or performant.

- With events related to time-series data or other tracking events that have a very high volume or rate.

- Following edits to a large **create, read, update, delete** (**CRUD**) resource. Instead of sending the entire updated resource, you might send a list of updated fields only.

When you use notifications, you are expecting the interested consumers to eventually make a call back to you to retrieve more information, as depicted in the following diagram:

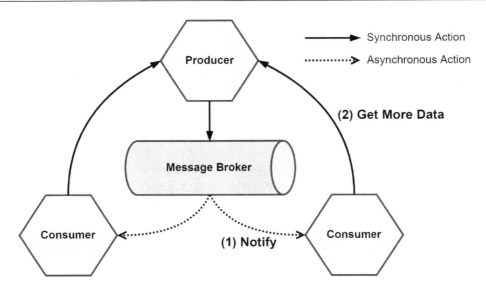

Figure 6.2 – Notifications and the resulting callbacks

The **Producer** in the preceding diagram will need to be scalable to handle the extra load from the callbacks—callbacks that are from potentially an unknown number of interested consumers. Compared to event-carried state transfer, notifications do not completely decouple the components.

The **Consumer** will need to also know where to find the **Producer** and should have implemented the **application programming interface (API)** to make a callback. Likewise, the **Producer** needs to have an API so that additional data can be retrieved. The **Consumer** is also temporally coupled to the **Producer**, so availability is still a risk, meaning if the **Producer** is down or not responding, then it is on the **Consumer** to handle the error and have the logic to retry fetching the data later when it is again available.

Between the **Producer** and **Consumer** sits the **Message Broker**, which contains the queues that the messages are published to and consumed from. The **Message Broker** does provide a level of decoupling between the **Producer** and **Consumer**, but because the **Consumer** makes calls back to the **Producer** for more information, the decoupling is not very strong.

Using notifications and callbacks to optimize network traffic may not always work out as planned. If a resource changes more rapidly than a consumer can consume an event and request information, data loss may result, as depicted in the following diagram:

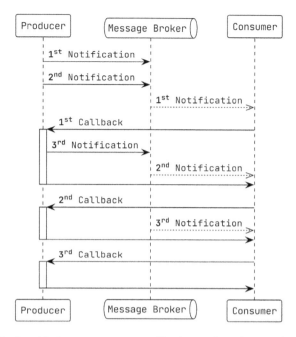

Figure 6.3 –The 2nd and 3rd are unnecessary callbacks resulting from multiple notifications

When the **Producer** in the preceding sequence diagram sends the second and third notifications, the latency to the **Consumer** creates a situation where multiple requests are being made for data that the **Consumer** already possesses.  Just as likely, the second or third callbacks could end being made before a previous one has finished. Solutions such as serializing and debouncing the callbacks could help with this situation.

## Integration with event-carried state transfer

Consumers are much less likely to need to make a request back to the producer for more information when they communicate with event-carried state transfer. **State transfer** is great for interested consumers to build a local representation of the data so that it may handle future requests independently.

Some uses for **event-carried state transfer** are presented here:

- Storing shipping addresses for customers in a warehouse service

- Building a history of product purchases for a seller in an independent search component

- Information from multiple producers can be combined to create entirely new resources for the application to support additional functionality

The primary advantage of event-carried state transfer is that consumers are temporally decoupled from the producers. Availability of the producer is no longer a factor in how resilient the consumer will be when it comes to handling requests that it receives.

Stateful events may contain only the data related to the specific change, or they can contain complete old and new representations of the resource, or a delta of a resource that was altered after the change was applied. A trap with stateful events is putting in too much data or trying to include information that is assumed to be useful for specific consumers. Finally, events should never contain information that was received from another domain. For example, an event coming from a sales domain should not include a shipping schedule that it received from the warehouse domain.

A balance on the amount of state is important, and so is the number of events that are being sent. Not every domain event is useful outside of the domain it sprang from. Consider the usefulness of the information and the event before creating a firehose that everyone is expected to consume to get a limited number of events that they are interested in.

Keeping a local copy is not without issues either. Information necessary to complete an operation could be missing because it has not arrived yet or a message was lost. Making a call or publishing a query message to the information owner to retrieve the data could be done to resolve the inconsistency. Putting the message back into the queue and retrying later may also work.

## Eventual consistency

**Eventual consistency** is constant in distributed applications and especially in event-driven applications. It is a trade-off made for the performance and resiliency gains when choosing to architect a system with asynchronous communication patterns.

Here's a quick definition of what eventual consistency is: *An eventually consistent system that has stopped receiving modifications to an item will eventually return the same last update across the system.*

It is a good chance that if you are working with microservices and are using synchronous communication patterns, then you are at least aware of and are somewhat comfortable with eventual consistency. If you are working with a monolith—even a modular monolith such as our little application—you might not be aware or comfortable with it.

Both kinds of integration events can result in an inconsistent system state. When an asynchronous system is operating normally, there can be no noticeable difference when compared to the synchronous equivalent. However, the additional infrastructure and complexity brought into the architecture add more places for errors to occur.

Eventual consistency is not always going to be a problem, or even present itself in catastrophic ways. When adding a new product to a store, the resulting change may take a little time to propagate through the system. If a customer were to call up the catalog for the store before the change arrived, they would not be affected by the inconsistency unless they were specifically aware and looking for the product.

Where eventual consistency can go wrong is when a state change is made and immediately, on returning a successful response to the client, a read is performed that attempts to read that state change, as depicted in the following diagram:

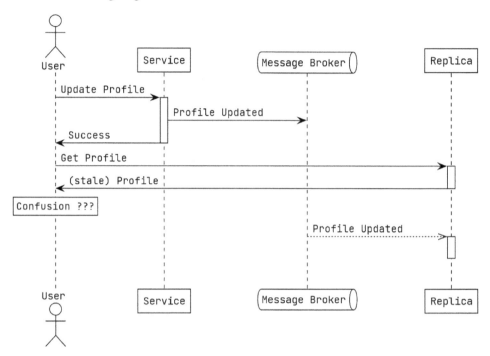

Figure 6.4 – Read-after-write inconsistency while updating the profile for a user

More than likely, the read will be sent to a different location or to a replica of where the write was made initially, and stale data is returned. This is called **read-after-write inconsistency**, and it has to do with not being able to read the state change or new data immediately after writing it.

One solution for the example from *Figure 6.4* would be to read from the primary database when a user requests their own profile. Any other user viewing a profile belonging to another user will not know they are not seeing the absolute latest update when viewing another user's profile. More solutions might be usable. Using a cache layer that is going to be updated more quickly might work, or the **user interface** (**UI**) that the user is using could not make the request for the updated profile at all and instead displays the information the user entered instead.

## Message-delivery guarantees

**Event-driven architectures** (**EDAs**) can be built around different levels of delivery guarantees. There are three possible options, and all three may be available depending on the broker or libraries you use.

### At-most-once message delivery

The **Producer** does not wait for an acknowledgment from the **Message Broker** when it publishes a message under the *at-most-once delivery* model, as depicted in the following diagram:

Figure 6.5 – At-most-once delivery

Message deduplication and idempotency are not a concern. However, the possibility the message never arrives is very real. In addition to the **Producer** not confirming that the **Message Broker** received the message, the broker does not wait for any acknowledgment from the **Consumer** before it deletes the message. If the **Consumer** fails to process the message, then the message will be lost.

At-most-once delivery guarantees can be put to good use in several situations, such as the collection of logs and processing messages from **Internet of Things (IoT)** devices.

### At-least-once message delivery

With *at-least-once delivery*, the **Producer** is guaranteed to have published the message to the **Message Broker**, and then the broker will keep delivering the message to the **Consumer** until the **Message Broker** has received an acknowledgment that the message has been received, as depicted in the following diagram:

Figure 6.6 – At-least-once delivery

A **Consumer** may receive the message more than once, and they must be utilizing either message deduplication or have implemented other idempotency measures to ensure that the redelivery of a message does not result in it being processed more than once.

The reasons why a message might be delivered more than once can vary, but it will often be because the **Message Broker** is waiting a limited amount of time for an acknowledgment from the **Consumer**. If the **Consumer** takes too long to send an acknowledgment, then the message is requeued to be sent again.

Systems that can deduplicate messages so that repeated deliveries only result in one processing instance are the ideal use case for at-least-once delivery.

### Exactly-once message delivery

Having a guarantee that a message will arrive exactly once is not so simple. As with the *at-least-once delivery* guarantee, the **Producer** will wait for an acknowledgment from the broker. Also, the broker will keep delivering the message until it has received an acknowledgment from the receiver, as depicted in the following diagram:

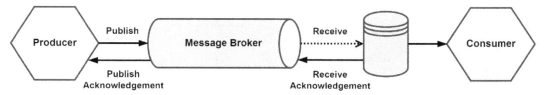

Figure 6.7 – Exactly-once delivery

What is different now is that what received the message was not the **Consumer** but instead an additional component that holds a copy of the message. The message can then be read, processed, and deleted by the **Consumer**. That is at least the idea of how *exactly-once delivery* can be achieved, but network reliability and issues with the **Message Broker** or with the message store can all still cause the process to fail.

*Exactly-once delivery* would be ideal for just about any situation, but it is extremely hard or downright impossible to achieve in most cases.

## Idempotent message delivery

Not every application will be able to deploy the infrastructure to have *exactly-once message delivery* and others will not need it. When most people think of *exactly-once delivery*, what comes to their mind is exactly-once processing of messages. This goal of exactly-once processing of messages can be achieved by adding deduplication to *at-least-once delivery*.

The most common technique is to deduplicate the receipt of the message using the identity of the message. Using the messaging library or middleware, the identity for the message is checked against a list of already received and processed message identities. If the identity already exists, then the message is acknowledged and discarded. If the identity is not found, then the request continues to message processing. The process is illustrated in the following diagram:

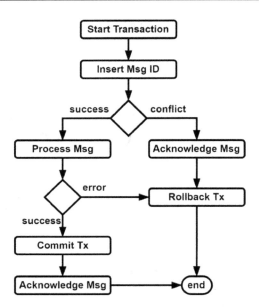

Figure 6.8 – Deduplication of incoming messages using transactions

A database transaction can be used to create a unit of work around the deduplication and the message processing. When the messaging processing fails, the transaction can be rolled back to remove the message identity from the database. When the message processing has succeeded, we make sure to commit the transaction before acknowledging the message with the message broker.

## Ordered message delivery

As with delivery guarantees, the order you will receive events comes with its own scale of guarantees of order. You can quickly find yourself in hot water if you listen to your vendor who promises that their product always delivers messages in the order they were published and later learn you are processing ProductRemoved events before the corresponding ProductAdded event.

The number of consumers you use and how you use them can have a huge impact on ordering.

- **Single consumer**: A single consumer subscribed to a **First-In, First-Out** (**FIFO**) queue will receive messages in the order that they were published, as depicted in the following diagram:

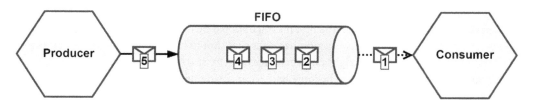

Figure 6.9 – Single consumer receiving messages in order from a FIFO queue

If our system were to publish messages at or below the rate it consumes them, then a single consumer will keep up and be all that we need. This is often not the case in an event-driven application.

- **Multiple consumers**: To handle higher volumes of messages, we can add additional consumers to keep the process rate steady. The additional consumers would be added to share the queue, essentially competing for the next message in the queue, and this is how the competing consumer pattern got its name. You can see a depiction of such a situation here:

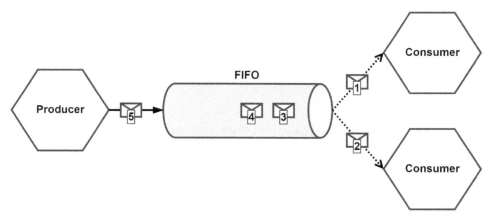

Figure 6.10 – Multiple consumers competing for messages from a FIFO queue

Having additional consumers will help with the rate at which messages can be processed and is a very common pattern. It does, however, create a potential issue with the order in which related messages will be processed.

In *Figure 6.10*, both consumers have received messages, and we will assume these messages belong to the same resource somewhere. In the queue, these messages were ordered, but now they are being processed concurrently. We cannot guess which consumer will finish first, and we may run into problems while processing the second message.

Starting with the least sophisticated solution first, we could let the second message be requeued to be delivered again. If timed correctly, then the first message should be done before the requeued message is delivered again to a consumer. This solution has risks that should be obvious. Requeuing messages forever, hoping that eventually, the right circumstances will exist so that they can be processed will deadlock your queue. The queue would become backed up, bringing any operations that were reliant on it to a standstill. You would eventually need to have messages that cannot be processed go elsewhere, such as a dead-letter queue. From there, it is purely situational how you want to proceed.

If the order of the messages causing problems are all related—say, because they belong to the same aggregate or the same workflow—then using a partitioned queue will help keep the messages in order when they are finally delivered. The following diagram provides an illustration of this:

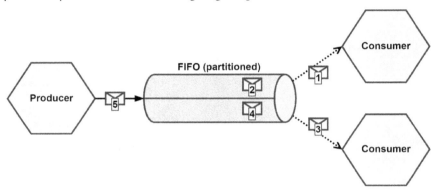

Figure 6.11 – Using partitions to maintain ordered delivery

With a partitioned queue, all messages with the same partition key will be delivered in the order that they were published for that partition. At most, a single consumer will be subscribed to any partition, and we conceptually return to the single-consumer example that we started with in this section. Each partition may have a single consumer subscribed to it, but the consumers may be subscribed to many partitions. When a queue is partitioned, it might be partitioned into 10, 25, or more partitions to allow for scaling the consumers in response to the load or for reliability.

> **Picking your partition keys**
>
> If you were to use partitions for the customers in your system, you would not have a partition per customer. Instead, you would have a partition for a subset of customers. You would provide the customer identity as the partition key, and the message broker would use it to compute which partition number to place the message in.

Even with using partitioned queues, you can get into trouble again if the consumer is processing messages asynchronously. So, in the Go language, if you were to process the messages using goroutines and the message broker was set up to deliver multiple messages or allow multiple messages to be inflight to a consumer, then you are right back where you were with the competing consumers, only now it would be competing goroutines.

Finally, sometimes processing messages out of order can be architected to not result in an inconsistent state. Take the `ProductPriceIncreased` and `ProductPriceDecreased` events as an example. Their payloads record a delta to the price. If these events were to arrive at some consumer, it would not matter which one was processed first because the state would eventually be consistent with the source.

## Implementing messaging with NATS JetStream

NATS (`https://nats.io`) is a very popular messaging broker that supports subject-based messaging and **publish-subscribe (pub-sub)**. Core NATS also supports load-balanced queue groups, so the competing consumer pattern can be used to scale up for higher message processing rates. It does not support, at least at the time of writing this book, partitioned queues.

NATS is capable of distributing millions of messages a second and, compared with many other message brokers, it has an easy-to-use API and message model, as described here:

- **Subject**: A string containing where the message is to be published or was published to.

- **Payload**: A byte slice capable of holding up to 64 **megabytes (MB)**; the NATS maintainers recommend smaller sizes, though.

- **Headers**: A map of string slices indexed with strings, not unlike the headers from the standard library `http` package.

- **Reply**: A string used to handle replying to an asynchronous request; we will not use the NATS **Request-Reply** feature because JetStream does not support it.

We will be making use of NATS JetStream, which is the replacement for NATS Streaming, an additional application that was used with NATS to provide durable streams. The NATS team wanted a better streaming experience and developed JetStream with goals such as a better user experience and a better management experience in mind. What they ended up with—JetStream—was not an additional application like NATS Streaming was but instead a part of the NATS application itself. To start NATS with JetStream enabled, we simply include the `-js` parameter to the command that we use to start the NATS server.

On top of what Core NATS provides, JetStream will provide durable streams and the ability to create NATS consumers with cursors to track their place on the stream. Our consumers will be able to subscribe to subjects and receive messages that were published well before the subscription was received by the server. There are a few more features JetStream adds to NATS Core that we are interested in, as outlined here:

- **Message deduplication**: This can deduplicate messages that have been published more than once

- **Message replay**: Consumers may receive all messages, or receive messages after a specific point in the stream or after a specific timestamp

- **Additional retention policies**: We can choose to keep messages if consumers exist with subscriptions to them or assign limits on the number of messages or total size of the stream

In the following diagram, we have shown how the JetStream components fit into an asynchronous message flow:

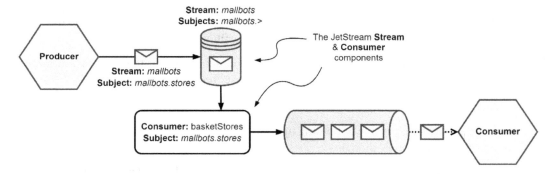

Figure 6.12 – NATS JetStream stream and consumer flow

JetStream provides two components, the **Stream** and the **Consumer**. They are described in more detail here:

- **Stream**: This is responsible for storing published messages for several subjects. Subjects may be named explicitly to be included or be included with the use of token wildcards. Message retention—based on duration, size, or interest—is configured independently for each stream. Our *MallBots* stream could be just one stream configured in JetStream alongside many others.

- **Consumer**: This is created as a *view* on the message store. Each consumer has a cursor that is used to iterate over the messages in a stream or a subset of them based on both a subject filter and replay policy.

We will use two packages to implement asynchronous communication in our application. These new packages will live under /internal and are set out here:

- The first is the am package. This will provide general asynchronous messaging functionality and interfaces.

- The second is the jetstream package, and it will provide **NATS JetStream**-specific functionality.

The way we will use these packages will be like how we used es, the event-sourcing package, and the postgres packages in the previous chapter.

## The am package

In the asynchronous package, we start with the message, as depicted here:

Figure 6.13 – The message and message handler interfaces

The Message interface is kept slim and focused on the management of a message that needs to be sent or received. Yes—event-driven applications communicate with events, but the event will not be the only message we will be communicating with. The MessageHandler interface is defined with a generic Message type, so we can avoid having to maintain handlers for every possible kind of message we will be using.

We want to be able to publish anything into a stream, so our MessagePublisher interface is going to need to be with a generic interface{ } or any type, as depicted in the following diagram:

```
                    <<Interface>>
                  MessagePublisher[T any]

    Publish(ctx context.Context, topicName string, v T) error
```

```
                    <<Interface>>
                MessageSubscriber[T Message]

    Subscribe(topicName string, handler MessagHandler[T], options ...SubscriberOption) error
```

```
                    <<Interface>>
              MessageStream[I any, O Message]

              MessagePublisher[I]
              MessageSubscriber[O]
```

Figure 6.14 – The message publisher, subscriber, and stream interfaces

For the MessageSubscriber interface, we will be returning a Message type of some kind, and so it has been defined to use the previously defined generic MessageHandler interface.

Finally, the MessagePublisher and MessageSubscriber interfaces are brought together into the MessageStream interface, which will allow us to create a stream that will let us publish an Event type and receive an EventMessage type. That is exactly what we do to create the EventStream type that we will be adding in this chapter, as illustrated in the following code snippet:

```
type EventStream = MessageStream[ddd.Event, EventMessage]
```

The am package will contain streams for the basic types of messages that we will be using—event, command, query, and reply—but the generics used in the interfaces shown earlier would permit even more types of messages should we need them.

For now, we will only be implementing an event stream, and the rest will be added in later chapters. For our event stream, we want to publish a ddd.Event type and to receive the EventMessage type. What we implement will need to serialize and deserialize events into something we can then pass into NATS JetStream, but we do not want to use a format specific for JetStream because that would create a dependency on NATS. The reason this would be bad is that it would then be more difficult to switch to different technologies and to test. For our intermediary type, we have the RawMessage interface and rawMessage struct, as depicted in the following diagram:

Figure 6.15 – The raw message intermediary interface and struct

With those last two components, we have what we need to create an `eventStream` struct that implements the `EventStream` interface, as shown here:

| eventStream |
| --- |
| reg registry.Registry<br>stream MessageStream[RawMessage, RawMessage] |
| **Publish**(ctx context.Context, topicName string, event ddd.Event) error<br>**Subscribe**(topicName string, handler MessageHandler[EventMessage], options ...SubscriberOptions) error |

Figure 6.16 – Our event stream implementation

Unpacking what is happening in the `eventStream` implementation, we have a `Publish()` method that accepts only the `ddd.Event` type and a `Subscribe()` method that only accepts handlers that operate on `EventMessages`. We need a registry to process the event payloads, and the event stream implementation will also need a stream that handles the `RawMessage` type for both the published input and the subscribed output types.

Another purpose for having an implementation for a specific message type is so that we can serialize and deserialize the data correctly. We could have also made a general `messageStream` and had the code doing the serialization work be passed in as a dependency. That might still happen, but while we only need a stream that handles events, we can avoid creating additional interfaces and the general implementation if we do not need it at this time.

The event stream `Publish()` method is primarily focused on event serialization work. You can see this in use in the following code snippet:

```
func (s eventStream) Publish(
    ctx context.Context, topicName string, event ddd.Event
        ) error {
    metadata, err := structpb.NewStruct(event.Metadata())
```

```
        if err != nil { return err }

        payload, err := s.reg.Serialize(
            event.EventName(), event.Payload(),
        )
        if err != nil { return err }

        data, err := proto.Marshal(&EventMessageData{
            Payload:    payload,
            OccurredAt: timestamppb.New(event.OccurredAt()),
            Metadata:   metadata,
        })
        if err != nil { return err }

        return s.stream.Publish(ctx, topicName, rawMessage{
            id:   event.ID(),
            name: event.EventName(),
            data: data,
        })
    }
```

We use a protocol buffer message as the data container that is then used as the data for the raw message. Here is the protocol buffer message that we use to serialize the event data with:

```
message EventMessageData {
  bytes payload = 1;
  google.protobuf.Timestamp occurred_at = 2;
  google.protobuf.Struct metadata = 3;
}
```

We only need to serialize the fields that do not go into the message. The payload is going to be taken care of by the registry. The OccurredAt and Metadata values for an event fit into the Timestamp and Struct known types respectfully.

The Subscribe() method does the same steps that the Publish() method does but in reverse. The outcome of running those steps in reverse goes into an instance of the eventMessage struct that has implemented both the ddd.Event and EventMessage interfaces. Together, these interfaces create an EventMessage interface, as depicted here:

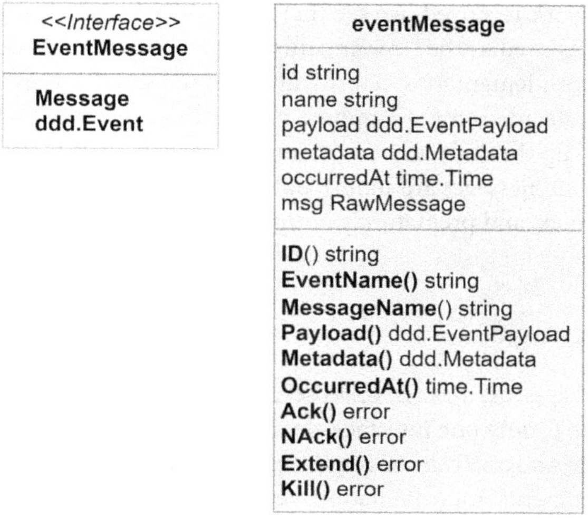

Figure 6.17 – The event message interface and struct

`Subscribe()` does a little more than just deserializing things. The deserialization work happens inside of a `MessageHandler` interface that it creates and passes into the raw message stream. The method is shown next, but with the already mentioned parts removed for brevity:

```
func (s eventStream) Subscribe(
    topicName string,
    handler MessageHandler[EventMessage],
    options ...SubscriberOption,
) error {
    fn := func(ctx context.Context, msg RawMessage) error {
        // ... eventMsg deserialization work
        return handler.HandleMessage(ctx, eventMsg)
    }

    return s.stream.Subscribe(
        topicName,
        MessageHandlerFunc[RawMessage](fn),
        options...,
    )
}
```

With the deserialization work removed, we see that the `Subscribe()` method only creates an anonymous function that is used as the `RawMessage MessageHandler` interface. The work that the `EventStream` implementation does is all about the serialization and deserialization of an event because we have decided to not combine it with the concerns of integrating with NATS JetStream or tried to DRY up the code and use a single stream handler for every possible message type we could imagine. Event messages are simple, but a future message-type stream implementation could be much more complex, and prematurely optimizing the implementations we create may not work out how we planned.

## The jetstream package

As with the `postgres` packages, the `jetstream` package holds infrastructure-specific implementations for NATS JetStream. There is only one interface we need to implement, and that is `RawMessage MessageStream`. The `MessageStream` implementation in the `jetstream` package is not that dissimilar to the `EventStream` implementation we looked at only a few pages back. It's described in more detail here:

*   `Publish()` is going to serialize the `RawMessage` into a NATS message.
*   `Subscribe()` again is doing the opposite within a handler function that is passed into either a JetStream `Subscribe()` or `QueueSubscribe()` method. The `QueueSubscribe()` method is used when you want to create a subscription with competing consumers.

### *Why do we have a jetstream package?*

We use packages for our infrastructure so that they are easy to swap out but also so that the nuances of having to deal with a specific infrastructure do not influence the design of our applications. With NATS JetStream, and PostgreSQL too, the work you—the reader—would need to put in to swap NATS out to try a different messaging broker, such as Apache Kafka or RabbitMQ, is not a heavy lift.

In the next section, we will update the application to begin publishing messages from the **Store Management** module, and you will get a clear idea of what will need to be changed if you wish to experiment with different message brokers.

## Making the Store Management module asynchronous

We are going to update the **Store Management** module to publish integration events and will also update the **Shopping Baskets** module to receive messages. The **Shopping Baskets** module will not be doing much more than logging the receipt of the message. Using the data will come in handy in the next chapter when we learn about event-carried state transfer.

## Modifying the monolith configuration

Starting with a simple configuration for NATS, we need a **Uniform Resource Locator (URL)** to connect to and a name for our stream. Of course, both could be hardcoded or put into variables in the code, but I run the application from a Docker container and without. The stream name is used in a few places, so having it be part of the configuration for the application is the lazy option. The code is illustrated here:

```
NatsConfig struct {
    URL    string `required:"true"`
    Stream string `default:"mallbots"`
}
```

The preceding code is added to the `AppConfig` with the field name `Nats`. To access the connection URL, we would use `cfg.Nats.URL`. For the Docker Compose environment, NATS will be available at `nats:4222`.

## Updating the monolith application

First, we need to update the monolith to prepare things for the modules, as follows:

- Modify the monolith configuration so that it can accept NATS JetStream settings
- Connect to NATS and add a graceful shutdown for the connection
- Update the monolith so that it will provide the modules with a `JetStreamContext` value

In the monolith application, `cmd/mallbots/monolith.go`, we need a field for the NATS connection and another one for the `JetStreamContext` value, as shown here:

```
type app struct {
    nc    *nats.Conn
    js    nats.JetStreamContext
    // ... other fields
}
```

In the composition root, we will connect to NATS, handle the error, and then create a `JetStreamContext` value. This is added in two parts. The initial connection will be in the monolith composition root, but the context creation will happen in a function, for organizational purposes and no other reason. The code is illustrated here:

```
m.nc, err = nats.Connect(cfg.Nats.URL)
if err != nil { return err }
```

```
defer m.nc.Close()
m.js, err = initJetStream(cfg.Nats, m.nc)
if err != nil { return err }
```

We will encapsulate the setup of the stream context and stream inside of the initJetStream() function, like so:

```
func initJetStream(
    cfg config.NatsConfig, nc *nats.Conn
) (nats.JetStreamContext, error) {
    js, err := nc.JetStream()
    if err != nil { return nil, err }
    _, err = js.AddStream(&nats.StreamConfig{
        Name:     cfg.Stream,
        Subjects: []string{
            fmt.Sprintf("%s.>", cfg.Stream),
        },
    })
    return js, err
}
```

In the first code block, a connection is made to NATS, and since we will not be leaving this function until we shut down the application, we include a deferred Close() call. In the initJetStream() function, we start by asking for a JetStreamContext value. If JetStream is not enabled for the server that we are connected to, then this would fail.

We then make a call to AddStream(), which will fail if we try to change the settings that were used for an already existing stream with the same name. The call is otherwise idempotent. However, if you do need to change the settings, then you will need to use UpdateStream(); then, be sure those new settings are used here in this call.

### *Gracefully shutting down the NATS connection*

As much as possible, we should do our best to shut down the application without losing any messages. To help with that, the NATS connection has a Drain() method that will unsubscribe all subscriptions and wait for any inflight messages to finish processing or be published before closing the connection. You can see an illustration of this in the following code snippet:

```
func (a *app) waitForStream(ctx context.Context) error {
    closed := make(chan struct{})
    a.nc.SetClosedHandler(func(*nats.Conn) {
```

```
            close(closed)
        })
        group, gCtx := errgroup.WithContext(ctx)
        group.Go(func() error {
            fmt.Println("message stream started")
            defer fmt.Println("message stream stopped")
            <-closed
            return nil
        })
        group.Go(func() error {
            <-gCtx.Done()
            return a.nc.Drain()
        })
        return group.Wait()
}
```

Here is what this method is doing:

- First, a channel is created that will be used as a semaphore to signal the connection has been fully closed.

- A handler is added to the NATS connection so that we can close the closed semaphore channel. The handler will be called after all the subscriptions and publishers have finished closing.

- An error group is created with the context that was provided to the method. The context that was passed into the method will cascade a cancelation or error down to the group context, allowing it to begin shutting down.

- In the first group function, there is not much going on besides outputting information to the console. The function will not exit on its own until the closed semaphore has been closed.

- In the second group function, we immediately wait for the group context to be canceled. After it is canceled, we will call Drain() on the NATS connection to gracefully shut down and begin closing the subscriptions and publishers.

- On the last line, the result of waiting for the error group is returned. This call blocks until all the group functions have exited.

## Providing to the modules the JetStreamContext

The `Monolith` interface in `internal/monolith/monolith.go` is updated with the `JS()` method so that modules can access the context, as illustrated here:

```
type Monolith interface {
    Config() config.AppConfig
    DB() *sql.DB
    JS() nats.JetStreamContext
    Logger() zerolog.Logger
    Mux() *chi.Mux
    RPC() *grpc.Server
    Waiter() waiter.Waiter
}
```

Then, the monolith application instance in `cmd/mallbots/monolith.go` is updated to implement the new method, like so:

```
func (a *app) JS() nats.JetStreamContext {
    return a.js
}
```

We may now use NATS JetStream in each module. Adding NATS JetStream to our application did take some work, but I would categorize it as more tedious than difficult.

> **Swapping out infrastructure**
>
> The monolith application modifications are the bits that would need to be altered if you were to swap out NATS for another message broker. The modules would use a different method on the monolith instance for the infrastructure-specific value and reference a different package for the stream implementation that works for the new infrastructure.

## Publishing messages from the Store Management module

The integration events we will be publishing from the **Store Management** module are going to be used by several other modules eventually, but in this chapter, only one module will be updated.

In real-world applications, we may not know how many consumers we have, and that is why integration events must be the most stable kind of event we have in our application. As I have stated before, if the event we are dealing with is only used by us and is never stored, we are free to change that event in any way we wish. So, we will not want to publish our domain events or the events we use for our event-sourced aggregates.

Each module exposes only its protocol buffer API, and that is where we will define all new integration events for the **Store Management** module.

We are going to follow a few rules on how we will be creating these events, as follows:

- The events need to be public, so all the events need to be defined in the `storespb` package.

- The events need to stand alone and not include any requests, responses, or other messages used by the **Google Remote Procedure Call** (**gRPC**) API.

- We do not want to expose how our module works, so that means we will not use the `AggregateEvent` type.

- Each event declaration must contain all the data we want to transport, and that includes identity references back to our models

### Defining our public events as protocol buffer messages

An `events.proto` file is used to help with organizing our integration events and to keep them separate from the **gRPC API** messages. When defining events you will be publishing, you want to avoid publishing so many events that the rest of the application will not be interested in, but in our little application, to make things easy, we will define an analog to the events we already have defined in the domains.

The `StoreCreated` and `StoreParticipationToggled` events as shown here as examples of how the messages will be constructed:

```
message StoreCreated {
  string id = 1;
  string name = 2;
  string location = 3;
}

message StoreParticipationToggled {
  string id = 1;
  bool participating = 2;
}
```

These two protocol buffer messages are very similar to the events we have defined in the domain, but it is important that we include a field for the store identity.

> **Duplicate event names**
>
> When we generate the Go code for these events, we will have domain events and integration events with the same name. Only the module that can see both would be aware of the names being duplicated. If using similarly named events seems like a problem, the integration events can of course be named differently.

## Making the events registerable

We want to make it easy for the consumers to use our events, and that means we need to add a bit of boilerplate code to the `storespb` package, as follows:

```go
const (
    StoreAggregateChannel = "mallbots.stores.events.Store"
    StoreCreatedEvent = "storesapi.StoreCreated"
    StoreParticipatingToggledEvent =
        "storesapi.StoreParticipatingToggled"
    // ... other constants
)


func Registrations(reg registry.Registry) (err error) {
    serde := serdes.NewProtoSerde(reg)
    err = serde.Register(&StoreCreated{})
    if err != nil { return err }
    err = serde.Register(&StoreParticipationToggled{})
    if err != nil { return err }
    // ... more registrations
    return nil
}
```

The code should define as constants the event keys that each payload uses for registration. The channels, called subjects in NATS, should also be defined as constants. An exported function should also be added so that any module can provide the registry instance it is using to have these events added to the registry.

In the preceding listing, we do the following:

- We define the channel for the `Store` aggregate event messages
- We define the key constants; not shown are the `Key()` implementations

- We have an exported `Registrations` function
- We register the protocol buffer events with `ProtoSerde`

## Updating the module composition root

The events we just added will need to be registered with the registry so that we may publish them. The function we added to register the events should be added either before or after domain event registrations.

The next addition we need to make to the composition root will be the code to create an event stream instance, as shown here:

```
eventStream := am.NewEventStream(
    reg,
    jetstream.NewStream(
        mono.Config().Nats.Stream,
        mono.JS(),
    ),
)
```

We now have an event stream ready to publish events and subscribe to subjects to receive event messages.

> **Minimal NATS JetStream presence**
>
> The monolith configuration changes and the method returning the `JetStreamContext` value will be used only in the composition root. If the message broker was swapped out, this is the only place that would need to be changed in the module.

## The concern of where to publish integration events from

We have a choice in front of us. We could pass an instance of the event stream into the application instance to publish the integration events directly from the commands. We could also create domain event handlers to act as a middleman between the application and the publication of the integration events. The trade-off being made is this: publishing directly from the commands may have access to information that will not be available from a domain event. Both approaches are valid, and in a different application, these may not even be the only two options.

## Adding integration event handlers

We will be working with events that are very similar to our domain events so that it will be easier to use handlers. We can always swap out a place or two if we need to—this is not an either/or choice.

Our integration event handlers will receive the event stream as a dependency. When we get a `StoreCreated` domain event, we will publish a new event with the event name and payload coming from the `storespb` package, as follows:

```
func (h IntegrationEventHandlers[T]) onStoreCreated(
    ctx context.Context, event ddd.AggregateEvent
) error {
    payload := event.Payload().(*domain.StoreCreated)
    return h.publisher.Publish(ctx,
        storespb.StoreAggregateChannel,
        ddd.NewEvent(storespb.StoreCreatedEvent,
            &storespb.StoreCreated{
                Id:       event.ID(),
                Name:     payload.Name,
                Location: payload.Location,
            },
        ),
    )
}
```

Had we chosen to publish from the application, it would be done essentially the same way. The important part is that we are publishing using constants and payload types that are available to the entire application.

### Finishing by connecting the handlers with the domain dispatcher

Back in the composition root, we can write this up with the logger as we have with the other event handlers, as follows:

```
integrationEventHandlers :=
    logging.LogEventHandlerAccess[ddd.AggregateEvent](
        application.NewIntegrationEventHandlers(eventStream),
        "IntegrationEvents", mono.Logger(),
    )
```

Then, finally, we connect the domain events dispatcher with our new handlers, like so:

```
func RegisterIntegrationEventHandlers[T ddd.AggregateEvent](
    eventHandlers ddd.EventHandler[T],
    domainSubscriber ddd.EventSubscriber[T],
```

```
) {
    domainSubscriber.Subscribe(eventHandlers,
        domain.StoreCreatedEvent,
        domain.StoreParticipationEnabledEvent,
        domain.StoreParticipationDisabledEvent,
        domain.StoreRebrandedEvent,
    )
}
```

The **Store Management** module is now set up to publish the first integration events. Next up is adding the receiving end in the **Shopping Baskets** module.

## Receiving messages in the Shopping Baskets module

To receive event messages, the initial composition root changes are very much the same, as outlined here:

- We need to register the `storespb` events with our registry
- We need to create an event stream instance

### *Adding store integration event handlers*

On the receiving side, we will always be using event handlers. The `EventHandler` instance we need to create is just like the domain and aggregate event handlers we have been working with in the past couple of chapters.

For now, we will log a debug message when we receive an event so that we can verify that we are really communicating with events. On this end, when we get a `StoreCreated` event, the event defined in the `storespb` package, it will have been serialized, sent over, and deserialized back into our event. The code is illustrated in the following snippet:

```
func (h StoreHandlers[T]) onStoreCreated(
    ctx context.Context, event ddd.Event
) error {
    payload := event.Payload().(*storespb.StoreCreated)
    h.logger.Debug().Msgf(
        `ID: %s, Name: "%s", Location: "%s"`,
        payload.GetId(),
        payload.GetName(),
        payload.GetLocation(),
    )
```

```
        return nil
    }
```

This `StoreHandlers` handler is not going to be any different from the other handlers in the **Shopping Baskets** module and can be set up with logging like the rest.

### Subscribing to the store aggregate channel

While we get to treat the integration event handlers on the receiving end no different, we do need to create a subscription a little differently. On the sending side, we subscribed like we have been and created a subscription with the domain dispatcher. Here, on the receiving end, we need to create a subscription on the event stream. It is done a little differently but there's nothing complicated, as we can see here:

```
func RegisterStoreHandlers(
    storeHandlers ddd.EventHandler[ddd.Event],
    stream am.EventSubscriber,
) error {
    evtMsgHandler :=
        am.MessageHandlerFunc[am.EventMessage](
        func(
            ctx context.Context,
            eventMsg am.EventMessage,
        ) error {
            return storeHandlers.HandleEvent(
                ctx,
                eventMsg,
            )
        },
    )

    return stream.Subscribe(
        storespb.StoreAggregateChannel,
        evtMsgHandler,
    )
}
```

`StoreHandlers` is an event handler; it has `HandleEvent` and not `HandleMessage`, and so it does not implement the method we need to receive the `EventMessage` type. The most type-safe way to get around this is to use the `MessageHandlerFunc` helper to wrap our handler so that it can receive the events it expects.

## Verifying we have good communication

Now, when we create a new store in the **Store Management** module through the **Swagger UI**, we will see something very much like this logged in the monolith container:

```
INF --> Stores.CreateStore
INF --> Stores.Mall.On(stores.StoreCreated)
INF <-- Stores.Mall.On(stores.StoreCreated)
INF --> Stores.IntegrationEvents.On(stores.StoreCreated)
INF <-- Stores.IntegrationEvents.On(stores.StoreCreated)
INF <-- Stores.CreateStore
INF --> Baskets.Store.On(storesapi.StoreCreated)
DBG ID: …, Name: "Waldorf Books", Location: "Upper Level
    West"
INF <-- Baskets.Store.On(storesapi.StoreCreated)
```

At the top of the log, we see the application call and then the two domain event handlers—one for the mall read model, with the other one being our new integration event handlers. After that, it shows the event has made its way to another module.

If the order of the log messages does not line up with the previous log, do not be alarmed by this. We are publishing the events asynchronously, in addition to them being asynchronous messages, so the **Shopping Baskets** module could receive the message before the `CreateStore` command has been completed in the **Store Management** module.

## Summary

In this chapter, we have finally achieved asynchronous communication. We covered the types of messages that are used in an event-driven application. We learned that events are messages, but messages are not always events. Messages have different kinds of delivery guarantees, and there are some important traps we need to be aware of when architecting an application with asynchronous communication patterns. NATS JetStream was introduced, and then we implemented an event stream using it as our message broker. We created integration events using protocol buffers and used the familiar event-handler patterns to both publish and receive these new types of events.

Our first asynchronous messages have been delivered from the **Store Management** module to the **Shopping Baskets** module.

In the next chapter, we will improve how we send and receive states across modules. We will create local caches of states shared between modules and begin to reduce the amount of coupling that the modules each have.

# 7
# Event-Carried State Transfer

In the previous chapter, we added NATS JetStream to our application as our message broker. We also add the ability to publish messages from the **Store Management** module and added message consumers to the **Shopping Baskets** module. For now, we are only logging the messages as they are consumed, and that will be changing in this chapter.

In this chapter, we will be looking at the data that each module shares with other modules; we will evaluate what data should continue to be shared with events and what data can be excluded. We will be adding a new API and taking advantage of the opportunity to refactor some module interactions.

Data from multiple modules will be brought together to create an entirely new read model. The new module will be an advanced order search and will bring together data from customers, stores, products, and, of course, orders.

We will be covering the following topics in this chapter:

- Refactoring to asynchronous communication
- Adding a new order search module
- Building read models from multiple sources

## Technical requirements

You will need to install or have installed the following software to run the application or to try the examples:

- The Go programming language version 1.18+
- Docker

The source code for the version of the application used in this chapter can be found at `https://github.com/PacktPublishing/Event-Driven-Architecture-in-Golang/tree/main/Chapter07`.

# Refactoring to asynchronous communication

In the last chapter, we published messages from the **Store Management** module to the **Shopping Baskets** module. We focused on creating the mechanisms between the modules and had only logged to the console when a message arrived. What we started in that chapter was adding new inputs and outputs to the modules:

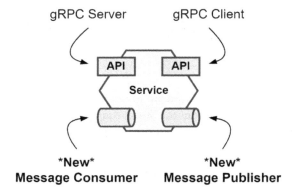

Figure 7.1 – New message inputs and outputs

We will be adding entirely new asynchronous APIs to the modules to implement the sharing of state via events: event-carried state transfer. Also, it would be an excellent time to reflect on the data each module is sharing with its existing gRPC API. We will be trying to determine what data other modules need to know about to function and where that data originates from.

## Store Management state transfer

The **Store Management** module shares `Store` and `Product` information with the other modules. It is the origin of all `Store` and `Product` data in our application. However, it is not the only module that shares that data with others. Here are the modules that use stores and products:

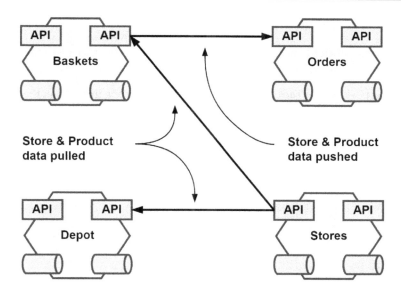

Figure 7.2 – Store Management data usage

The `Store` and `Product` data flows out from the **Store Management** module to the rest of the application. It is sometimes pulled, and sometimes it is pushed:

- **Shopping Baskets** and the **Depot** module make calls to pull in `Store` and `Product` data.
- **Order Processing** accepts `Store` and `Product` data pushed from the **Shopping Baskets** module in its `CreateOrder` endpoint.

The data that is being pulled into the **Shopping Baskets** and **Depot** modules could be replaced with local cached copies of the data. The data that is shared with the **Order Processing** module is secondhand data not owned by the calling module. Stores and products are used in **Order Processing** only when details about an order are being requested. We will make the following changes to the **Shopping Baskets** and **Depot** modules:

- Update the existing repositories to process data updates for stores and products
- Create new tables that will work as our local cache
- Use the existing gRPC calls as fallbacks when the local cache is missing data
- Update the integration event handlers to use the cache repositories

We will be leaving the `CreateOrder` call as-is for now, and we will visit that call when we are working on the workflow updates for the **Order Processing** module in the next chapter.

## Local cache for Stores and Products

The **Shopping Baskets** and **Depot** modules already define repository interfaces for **Store** and **Product** models that need to be updated to insert new rows and make updates when events come in.

Referring to these repositories as *cache repositories* may give the wrong impression that these should be temporary copies. Instead, I am intentionally adding the word *cache* so that for the demonstration it will be clear that this data is not changing owners when it is transferred between the modules. When the structure of either Product or Store changes, we may need to update the code that receives the event, but the rest of our module should remain unaffected. That receiving code will be acting as an anti-corruption layer, protecting our module from the external concerns of the **Store Management** module.

| *<<Interface>>* **ProductCacheRepository** |
|---|
| **Add**(ctx context.Context, productID, storeID, name string, price float64) error<br>**Rebrand**(ctx context.Context, productID, name string) error<br>**UpdatePrice**(ctx context.Context, productID string, price float64) error<br>**Remove**(ctx context.Context, productID string) error<br>**ProductRepository** |

| *<<Interface>>* **StoreCacheRepository** |
|---|
| **Add**(ctx context.Context, storeID, name string) error<br>**Rename**(ctx context.Context, storeID, name string) error<br>**StoreRepository** |

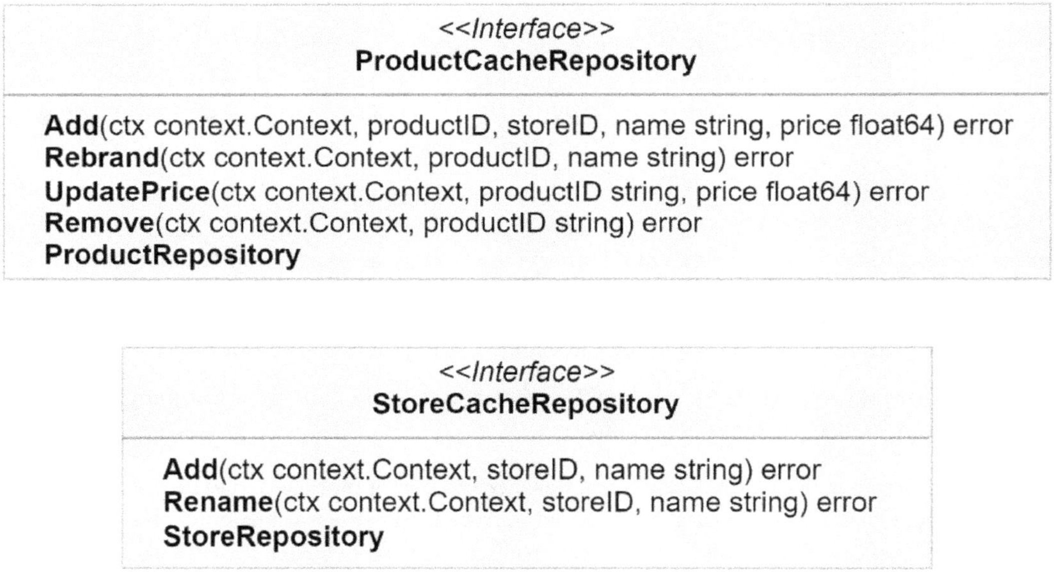

Figure 7.3 – Local cache interfaces for the Shopping Baskets module

The interfaces in *Figure 7.3* are similar to the ones we used for the MallRepository and CatalogRepository read models in the **Store Management** module. There are some minor modifications because we will not be interested in as much data in the local caches. The new cache repository interfaces can be used in place of the current repositories without any changes to any constructor that received them.

## Synchronous state fallbacks

We do not have to have the fallbacks, but since we already have the gRPC endpoints for the `Store` and `Product` data, we can choose to use those as fallbacks when we do not locate the requested data locally. This will not help us to determine if our cache is stale, and we will need to be careful about how we handle inserting new rows when they may already exist.

The Postgres implementations of the cache repository interfaces will accept a fallback parameter. When the data cannot be located locally, we will retrieve it from the fallback and then make a cached copy. We will implement the `Find()` method in a way that will use the fallback only if the error we get back from the database signifies that no rows were found:

```go
func (r StoreCacheRepository) Find(
    ctx context.Context, storeID string,
) (*domain.Store, error) {
    const query = "SELECT name FROM %s WHERE id = $1 LIMIT 1"
    store := &domain.Store{
        ID: storeID,
    }
    err := r.db.QueryRowContext(
        ctx, r.table(query), storeID,
    ).Scan(&store.Name)
    if err != nil {
        if !errors.Is(err, sql.ErrNoRows) {
            return nil, errors.Wrap(err, "scanning store")
        }
        store, err = r.fallback.Find(ctx, storeID)
        if err != nil {
            return nil, errors.Wrap(
                err, "store fallback failed"
            )
        }
        return store, r.Add(ctx, store.ID, store.Name)
    }
    return store, nil
}
```

Then, in the `Add()` method we will ignore errors that have to do with unique constraint violations. The reason we ignore these errors is that there may be a race to insert the data from an incoming message and from the gRPC fallback:

```
func (r StoreCacheRepository) Add(
    ctx context.Context, storeID, name string,
) error {
    const query = "INSERT INTO %s (id, name) VALUES ($1, $2)"
    _, err := r.db.ExecContext(
        ctx, r.table(query), storeID, name,
    )
    if err != nil {
        var pgErr *pgconn.PgError
        if errors.As(err, &pgErr) {
            if pgErr.Code == pgerrcode.UniqueViolation {
                return nil
            }
        }
    }
    return err
}
```

The rest of the methods in the cache repository implementations are very similar to their counterparts in the `MallRepository` and `CatalogRepository` implementations from **Store Management** module.

In the composition root for the **Shopping Baskets** module, we will change how the `stores` and `products` repositories are instantiated:

```
stores := postgres.NewStoreCacheRepository(
    "baskets.stores",
    mono.DB(),
    grpc.NewStoreRepository(conn),
)
products := postgres.NewProductCacheRepository(
    "baskets.products",
    mono.DB(),
    grpc.NewProductRepository(conn),
)
```

The new cache repositories should now replace the logging we used in the last chapter for the integration event handlers:

```
storeHandlers := logging.LogEventHandlerAccess[ddd.Event]
(     application.NewStoreHandlers(stores),
    "Store", mono.Logger(),
)
productHandlers := logging.LogEventHandlerAccess[ddd.Event]
(     application.NewProductHandlers(products),
    "Product", mono.Logger(),
)
```

The handlers will now use the repositories to add and update the cached data instead of logging the receipt of the various events, for example, when handling the `StoreCreatedEvent`:

```
func (h StoreHandlers[T]) onStoreCreated(
    ctx context.Context, event ddd.Event,
) error {
    payload := event.Payload().(*storespb.StoreCreated)
    return h.cache.Add(
        ctx, payload.GetId(), payload.GetName(),
    )
}
```

The **Shopping Baskets** module will now consume the events that the **Store Management** module is publishing to create a local cache that will make it more independent should either module be broken out of the monolith and made into a standalone microservice.

We can modify the **Depot** module in the same way to save a local cache of the data that it needs from the **Store Management** module. When we do implement it for that module, we would again look at what data it specifically needs and customize the cache that is implemented to support the right models:

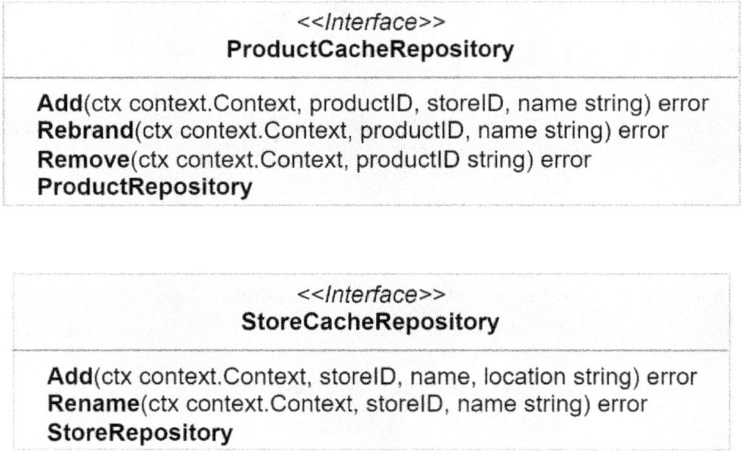

```
                          <<Interface>>
                      ProductCacheRepository

Add(ctx context.Context, productID, storeID, name string) error
Rebrand(ctx context.Context, productID, name string) error
Remove(ctx context.Context, productID string) error
ProductRepository
```

```
                          <<Interface>>
                       StoreCacheRepository

Add(ctx context.Context, storeID, name, location string) error
Rename(ctx context.Context, storeID, name string) error
StoreRepository
```

Figure 7.4 – The local cache interfaces for the Depot module

In the **Depot** module, the price of a product is not important and will not be stored when the product is created, nor will there be a need to record any price changes. In the store cache, the location of a store is important, and it will be cached.

Adding a store with **Swagger UI** will produce logs such as the following:

```
INF --> Stores.CreateStore
INF --> Stores.Mall.On(stores.StoreCreated)
INF <-- Stores.Mall.On(stores.StoreCreated)
INF --> Stores.IntegrationEvents.On(stores.StoreCreated)
INF <-- Stores.IntegrationEvents.On(stores.StoreCreated)
INF <-- Stores.CreateStore
INF --> Baskets.Store.On(storesapi.StoreCreated)
INF --> Depot.Store.On(storesapi.StoreCreated)
INF <-- Baskets.Store.On(storesapi.StoreCreated)
INF <-- Depot.Store.On(storesapi.StoreCreated)
```

Nothing has changed how the read model for the `Mall` is being processed: they will still be handled before the `Stores.CreateStore` command completes.

A second in-process handler for the integration events is run, which has asynchronously published the `storesapi.StoreCreated` event. Eventually, the two new consumers will receive the message and process it to create a cached copy of the store data. The order of the last four lines of the log will change depending on the speed at which each consumer can receive and process the message.

## Customer state transfer

The **Customer** module has not been the focus of much reworking in the past but, like **Store Management**, it maintains a resource that is of interest to the other modules: the customer data.

### Transferring the state but not the responsibility

A quick word of caution about the customer and other shared state in an event-driven application: when we transfer state with events, we do not transfer domain responsibilities along with it.

If the module or service that owns the customer data is also responsible for authorization or authentication, that responsibility stays with it, and it must continue to be called to perform that function.

Figure 7.5 – Customer data usage

Right now, as shown in *Figure 7.5*, only the **Notifications** module is using customer data. When the **Notifications** module receives a request to send out a notification, it needs to fetch the SMS number for the customer because the calling service is only able to pass along the customer's identity.

The following interface is what is used to create a local cache of customer data in the **Notifications** module:

<<*Interface*>>
**CustomerCacheRepository**

**Add**(ctx context.Context, customerID, name, smsNumber string) error
**UpdateSmsNumber**(ctx context.Context, customerID, smsNumber string) error
**CustomerRepository**

Figure 7.6 – The local cache interface for the Notifications module

The **Customers** module will need to be updated to become a publisher of integration events, and the steps to accomplish that can be found in the previous chapter in the *Publishing messages from the Store Management module* section.

Then, the **Notifications** module will need to be updated to receive those events, and the steps to do that are also described in the previous chapter, in the *Receiving messages in the Shopping Baskets module* section.

Requesting **Swagger UI** to create a new customer will produce results such as the following in the monolith logs:

```
INF --> Customers.RegisterCustomer
INF --> Customers.IntegrationEvents.On(customers.
CustomerRegistered)
INF <-- Customers.IntegrationEvents.On(customers.
CustomerRegistered)
INF <-- Customers.RegisterCustomer
INF --> Notifications.Customer.On(customersapi.
CustomerRegistered)
INF <-- Notifications.Customer.On(customersapi.
CustomerRegistered)
```

The order in which the logs appear may not be the same because the publishing and processing of the event will be asynchronous. We can see in these logs that the `Customers.RegisterCustomer` command is complete before the **Notifications** module starts to work on the message. If other modules also consume the messages published by the **Customers** module, we would not need to involve the team responsible to make it happen.

## Order processing state transfer

In *Chapter 4, Event Foundations*, we refactored the **Order Processing** module, extracting side effects from the command handlers into domain event handlers. One of those domain event handlers was for the notifications we wanted to send out when specific changes were made to the `Order` aggregate:

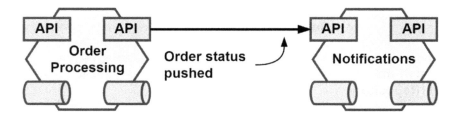

Figure 7.7 – The notification requests sent from the Order Processing module

Replacing the calls from the **Order Processing** module will not result in us creating a data cache in the **Notifications** module. What will happen instead is a reaction to the state change resulting in a notification being sent to the customer.

We will replace NotificationHandlers in the **Order Processing** module with a new IntegrationEventHandlers. After we do this, we will have completed an event refactoring journey and will have completely decoupled **Order Processing** from **Notifications**.

This is the handler for the domain event OrderReadied:

```
func (h IntegrationEventHandlers[T]) onOrderReadied(
    ctx context.Context, event ddd.AggregateEvent,
) error {
    payload := event.Payload().(*domain.OrderReadied)
    return h.publisher.Publish(
        ctx,
        orderingpb.OrderAggregateChannel,
        ddd.NewEvent(
            orderingpb.OrderReadiedEvent,
            &orderingpb.OrderReadied{
                Id:         event.AggregateID(),
                CustomerId: payload.CustomerID,
                PaymentId:  payload.PaymentID,
                Total:      payload.Total,
            },
        ),
    )
}
```

The payload we will be publishing is not as slim as the gRPC request to the **Notification** module, and that is because we will also want to use this to handle the other side effect that deals with creating invoices in the **Payments** module.

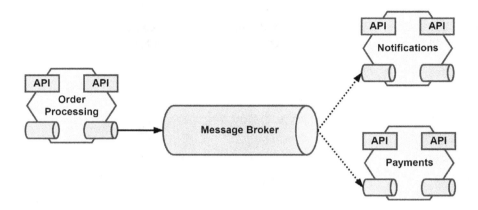

Figure 7.8 – Replacing side effect handlers with asynchronous messaging

After also updating the two modules, we will have removed them both as dependencies for the **Order Processing** module. There are still other dependencies on other modules, and we will be getting to them in the next chapter, when we update **Order Processing** to use asynchronous workflows.

### Other refactoring considerations

We may be able to or want to remove or deprecate the gRPC endpoints that were used to send the customer notifications now that we have a new asynchronous messaging alternative. Whether you should or how to handle the removal will be extremely situational and will require at the very least a survey of the API users to see if they can support the switch to the new asynchronous communication methods.

## Payments state transfer

The last state we want to update belongs to the **Payments** module, and it goes to the **Order Processing** module. When an invoice is paid, we want to update the order to put it into a final completed state:

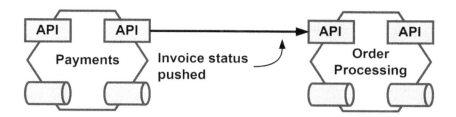

Figure 7.9 – Invoice status is pushed to Order Processing to complete orders

Replacing the call to **Order Processing** will remove the only dependency **Payments** had on other modules.

When the **Order Processing** module consumes the `paymentsapi.InvoicePaid` event it will kick off the same application task that it had before when the gRPC request was received.

## Documenting the asynchronous API

One of the advantages of building an event-driven application is that there is a decoupling between the producers of the events and the consumers. The only thing that teams need to do in order to get things done is consume the messages that are relevant to them, and they may do this without having to engage with or affect the timeline of the publishing team.

You could take this to mean that consumers who are interested in what you are publishing will be interested enough to crawl through your source code to figure out what is being published so that they can subscribe to it. I cannot speak for others, but when it comes to my plans to integrate components, that has nothing to do with any possible interest I might have.

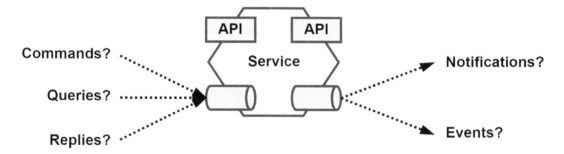

Figure 7.10 – Unknown asynchronous messaging landscape

The alternative is to, of course, maintain documentation for the asynchronous API. The organization could use a shared document or a wiki, but the issue with either of these options is the organization would also need to come up with what and how things need to be documented.

This is not a problem for some, but it does present an additional challenge, and bad documentation is often no better than no documentation.

## AsyncAPI

Providing a structured specification is exactly what **AsyncAPI** (`https://asyncapi.com`) is designed to do. It uses a specification schema that is very similar to the **OpenAPI** specification schema. Where OpenAPI would document the paths or endpoints and verbs that an API provided, AsyncAPI documents the channels and messages that a component would publish or subscribe to.

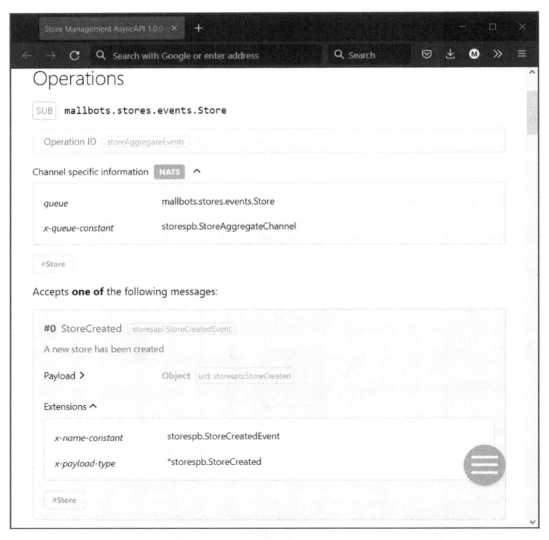

Figure 7.11 – AsyncAPI documentation for the Store Management module

The HTML documentation shown here was created using the AsyncAPI generator tool. The generator can also be used to generate boilerplate code for multiple languages or documentation as a PDF or in Markdown instead of HTML if preferred.

In the documentation generated for the modules, we include references using specification extensions for the constants and types in the Go code to reduce the need to get into the source code, unless they are interested in doing that.

## EventCatalog

Another promising tool is **EventCatalog** (`https://eventcatalog.dev`), which uses Markdown files and functions just like a static site generator.

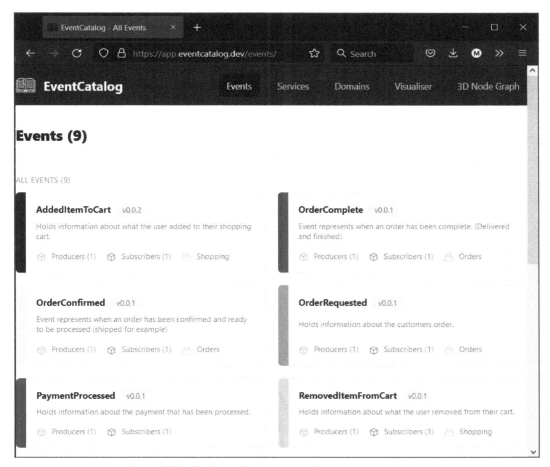

Figure 7.12 – The EventCatalog demo showing the events tab

In addition to being able to define the events and asynchronous APIs, you can also include documentation for the synchronous APIs for services as well. The generated site can provide a visualization of the relationships that services have through their events. The site can even render a 3D node graph of the entire system with animations showing the direction in which state flows.

With the knowledge that there are tools and specifications to document an event-driven application, there is no excuse to not document your asynchronous messaging APIs just like you would a REST or gRPC API.

## Adding a new order search module

Now that we are publishing the application state as it changes, we can consider new functionality that might have been impossible before or would have been too dependent on others to be feasible.

We will be consuming many different sources to keep a local cache to provide greater detail for our search results. Customer, store, and product names will all be stored locally. The new module will be consuming every message that the **Order Processing** module will be publishing to keep results current. Other data could also be included later, such as the status of the invoice, or the status of the shopping that takes place after the order has been submitted.

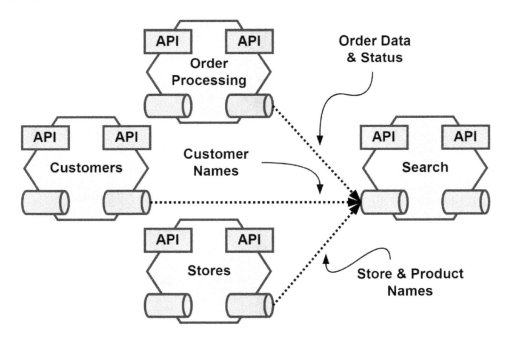

Figure 7.13 – The data that feeds the Search module

We will create the new module in a new directory called /search, and in that directory, we will create the module.go file exactly like the other modules. This new module will need the following driven adapters in the composition root:

- A data type registry instance
- Events from the **Customers**, **Store Management**, and **Order Processing** modules registered
- An event stream instance
- Several cache repositories with gRPC fallbacks
- The repository for our order read models

The following listing shows the registry instantiated as reg. eventStream uses the NATS JetStream implementation as the RawMessage source stream. The repositories for customers, stores, products, and orders are implemented using PostgreSQL:

```
reg := registry.New()
err = orderingpb.Registrations(reg)
if err != nil { return err }
err = customerspb.Registrations(reg)
if err != nil { return err }
err = storespb.Registrations(reg)
if err != nil { return err }
eventStream := am.NewEventStream(
    reg, jetstream.NewStream(
        mono.Config().Nats.Stream, mono.JS(),
    ),
)
conn, _ := grpc.Dial(ctx, mono.Config().Rpc.Address())
customers := postgres.NewCustomerCacheRepository(
    "search.customers_cache",
    mono.DB(),
    grpc.NewCustomerRepository(conn),
)
stores := postgres.NewStoreCacheRepository(
    "search.stores_cache",
    mono.DB(),
    grpc.NewStoreRepository(conn),
)
```

```
products := postgres.NewProductCacheRepository(
    "search.products_cache",
    mono.DB(),
    grpc.NewProductRepository(conn),
)
orders := postgres.NewOrderRepository(
    "search.orders",
    mono.DB(),
)
```

The application components will be as follows:

- An application with some query methods

- Three event handlers will create local caches of customer, store, and product data

- An order event handler to track the changes that are made as they happen

The dependencies from the previous listing are then injected into the application and handlers:

```
app := logging.LogApplicationAccess(
    application.New(orders),
    mono.Logger(),
)
orderHandlers := logging.LogEventHandlerAccess[ddd.Event](
    application.NewOrderHandlers(
        orders, customers, stores, products,
    ),
    "Order", mono.Logger(),
)
customerHandlers := logging.LogEventHandlerAccess[ddd.Event](
    application.NewCustomerHandlers(customers),
    "Customer", mono.Logger(),
)
storeHandlers := logging.LogEventHandlerAccess[ddd.Event](
    application.NewStoreHandlers(stores),
    "Store", mono.Logger(),
)
productHandlers := logging.LogEventHandlerAccess[ddd.Event](
```

```
        application.NewProductHandlers(products),
        "Product", mono.Logger(),
    )
```

Much of the work that this new module will do will all happen inside the handlers. The application has only two methods: `SearchOrders()` and `GetOrder()`. Because the handlers will consume events as the only form of input to produce the read models, the application will only need to have the two methods to perform queries.

For now, the handlers can function independently and work directly with the repositories. It is a design decision to not create application methods that are then used in the handlers, and it can be easily reversed if that would improve the maintainability of the module. The alternative is to add the methods to the application, which would result in our handlers functioning in the same way as the gRPC server methods would. The incoming message would essentially be transformed into application inputs by our handlers, and then they would process any errors that were returned.

Then, into the driver adapters, we inject the driven adapters, application, and handlers from the previous two listings:

```
err = grpc.RegisterServer(ctx, app, mono.RPC())
if err != nil { return err }
err = rest.RegisterGateway(
    ctx, mono.Mux(), mono.Config().Rpc.Address(),
)
if err != nil { return err }
err = handlers.RegisterOrderHandlers(
    orderHandlers, eventStream,
)
if err != nil { return err }
err = handlers.RegisterCustomerHandlers(
    customerHandlers, eventStream,
)
if err != nil { return err }
err = handlers.RegisterStoreHandlers(
    storeHandlers, eventStream,
)
if err != nil { return err }
err = handlers.RegisterProductHandlers(
```

```
        productHandlers, eventStream,
    )
    if err != nil { return err }
```

Much of what has gone into the composition root for the **Search** module is familiar to us at this point. The repositories, gRPC server, and REST gateway are also going to be standard and, aside from some changes to make them work locally, are copies of the ones we would find in the other modules. A large portion of this new module can be found or exists elsewhere in the other modules.

With that said, two questions spring to mind. Why create a new **Search** module, and why not make it part of the **Order Processing** module? The duties of handling order life cycles and doing complex searches on orders might have order data in common but the functionality does not entirely align. In a real-world application, we would not be dealing with such simple components, and adding entirely new functionality could introduce unexpected bugs or have other undesirable issues, such as reduced performance.

To answer the first question, this functionality also does not fit in with any other existing module. Plus, as has been stated many times by now, we can stand up a new component that consumes events in an event-driven application very easily. This new search feature and other functionality like it can be developed and vetted without causing any interruptions to other teams and developers, both in terms of scheduling pull requests to integrate the components and to development schedules.

## Building read models from multiple sources

The new **Search** module will be returning order data that should not require any additional queries to other services to be useful. We want to be able to return the customer's name, product name, and store names in the details we return. We also want to be able to locate the orders using more than their identities.

The search goals of this new module are as follows:

- Search for orders belonging to specific customer identities
- Search for orders by store and product identities
- Search for orders created within a date range
- Search for orders that have a total within a range
- Search for orders by their status

The read model that we will be building is not too different from what an order in the **Order Processing** module looks like:

| Order | |
| --- | --- |
| OrderID | string |
| CustomerID | string |
| CustomerName | string |
| Items | []Item |
| Total | float64 |
| Status | string |
| CreatedAt | time.Time |

| Item | |
| --- | --- |
| ProductID | string |
| StoreID | string |
| ProductName | string |
| StoreName | string |
| Price | float64 |
| Quantity | int |

Figure 7.14 – The order read model structures

To support searching using the previously mentioned filters, we will be writing some additional metadata along with the read model data we save. However, instead of making that additional data part of the read model structs, I prefer to have it live either alongside or in the PostgreSQL implementation of the `OrderRepository` interface. This kind of decision can save you a lot of time or headaches down the road if the current choice of database is unable to handle the load or support new methods of filtering.

We will be using PostgreSQL, and we will have the following as our table schema:

```
CREATE TABLE search.orders (
    order_id        text NOT NULL,
    customer_id     text NOT NULL,
    customer_name   text NOT NULL,
    payment_id      text NOT NULL,
    items           bytea NOT NULL,
    status          text NOT NULL,
    product_ids     text ARRAY NOT NULL,
    store_ids       text ARRAY NOT NULL,
    created_at      timestamptz NOT NULL DEFAULT NOW(),
    updated_at      timestamptz NOT NULL DEFAULT NOW(),
    PRIMARY KEY (order_id)
);
```

The `product_ids` and `store_ids` columns will make it easier to perform searches for orders that have been for specific products and stores.

> **Column data types**
>
> For our identifier columns, we could use the PostgreSQL *UUID* type and for the status column we could use the *enum* type. As this is a table that will be built using data from external sources, we would need to be careful that changes to the incoming data types does not cause problems. The use of an anti-corruption layer would help with that. The option I have chosen for the demonstration is to use data types for the columns that will have the fewest issues should the incoming data types change.

The rest of the search filters should not be hard, and we can improve performance with indexes. While the metadata and index additions are minimal, they would still be different for a different database.

## Creating a read model record

When the `OrderCreated` event is received, the data from that event is brought together with the data we have been storing with the other handlers. The previously mentioned metadata will be added just before saving the record into the database.

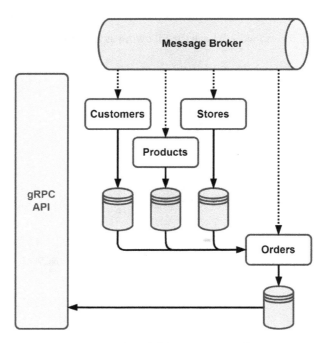

Figure 7.15 – Read model data sources and creation

Customer, store, and product data will be stored as it comes in. Later, when an order is created, we combine it with the data already stored in the database, creating our rich search model.

After the read model has been created, it will receive additional updates as the status changes. It will eventually be updated with the status. Our read model is going to be eventually consistent, and, under normal conditions, no one may ever notice.

## Summary

In this chapter, we used event-carried state transfer to decouple modules. Modules such as **Store Management** and **Customers** were made into event producers to improve the independence of the modules by allowing them to use locally cached data, avoiding a blocking gRPC call to retrieve it. We also expanded the state that is being shared in the application. Asynchronous messaging APIs can and should be documented like synchronous APIs, and we were introduced to a couple of tools that make the task easier.

We also added a new module to add advanced search capabilities to the application. This new module utilized events from several other modules to build a new read model that can be queried in multiple different ways.

We still have some synchronous calls that we did not touch. These calls will be the focus of our next chapter, *Chapter 8, Message Workflows*. In that chapter, we will look at how we can send more than events, and we will send commands to other modules so that they work at our behest. We will also look to address issues regarding lost messages and what can be done to prevent message loss in a busy application.

# Message Workflows

In the previous chapter, we used events to share information between the modules. In this chapter, we will learn how complex work can be done in a distributed and asynchronous way. We will introduce several different options for performing complex operations across different components. After that, we will implement a new asynchronous workflow for creating orders in the application using one of those techniques.

In this chapter, we will cover the following topics:

- What is a distributed transaction?
- Comparing various methods of distributed transactions
- Implementing distributed transactions with Sagas
- Converting the order creation process to use a Saga

## Technical requirements

You will need to install or have installed the following software to run the application or to try the examples in this chapter:

- The Go programming language, version 1.18+
- Docker

The code for this chapter can be found at `https://github.com/PacktPublishing/Event-Driven-Architecture-in-Golang/tree/main/Chapter08`.

# What is a distributed transaction?

The distributed components of an application will not always be able to complete a task completely isolated. We have already seen how we can use messages to share information between components so that remote components can have the data they need to complete small tasks. Within a simple component, more complex tasks could utilize a transaction to ensure that the entire operation completes atomically.

Let's talk about local transactions for a moment and why we would want to emulate them as distributed transactions. We use transactions for the **atomicity, consistency, isolation, and durability (ACID)** guarantees they provide us:

- The **atomicity** guarantee ensures that the group of queries is treated as a single unit – that is, a single interaction with the database – and that they all either succeed together or fail together

- The **consistency** guarantee ensures that the queries transition the state in the database while abiding by all rules, constraints, and triggers that exist in the database

- The **isolation** guarantee ensures that no other concurrent interactions with the database will affect this interaction with the database

- The **durability** guarantee ensures that once the transaction has been committed, any state changes made by the transaction will survive a system crash

Within a monolithic application, we may start a local transaction in the database so that all the interactions with the database use a singular view. We can also insert new data that will be atomically written together. Hypothetically, the following diagram shows what the create order process for MallBots would look like without modules and a single database:

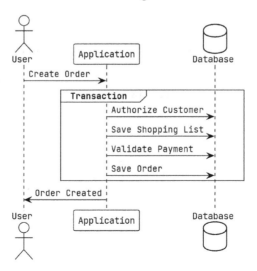

Figure 8.1 – Using a local transaction to create a new order

We are looking for the same ACID guarantees in the processes we use for distributed transactions. A distributed transaction should provide all or most of the guarantees that a local transaction would.

The only component necessary for a local transaction to be executed is either a **relational database management system** (**RDBMS**) or a non-relational (NoSQL) database that complies with ACID standards. Whereas a distributed transaction may include the entire system and perhaps several different kinds of databases, it is not limited to just one. Even a third-party service can be part of a distributed transaction. This brings us to another distinction between a local transaction and a distributed one: a distributed transaction has the potential to be run over a longer period. Also, some distributed transaction choices do not maintain the isolation guarantee so that resources are not blocked and are not fully ACID compliant:

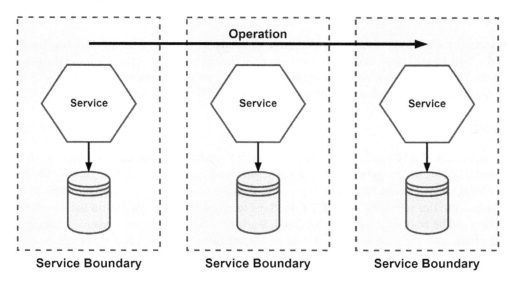

Figure 8.2 – An operation that runs across several services

An oversimplified example would be transferring money between two accounts. In one account, the money needs to be deducted, while in the other, the money is deposited. This operation can only be considered complete if both modifications are successful. If one of them fails, the modification must be undone on the other account.

## Why do we need distributed transactions?

Complex applications will inevitably have complex operations that cannot be contained to a simple component. Having that operation span the application without any way to keep the system consistent would be foolish at best.

Using *Figure 8.2* for another example, we have an operation that requires involvement from three different services. If we were to blindly pass on the operation to the second and third operations without any way to roll back a change in the previous services, our system as a whole could experience any number of issues. For example, the inventory could vanish, rooms could be reserved but left unbilled, or worse, the payment could have been accepted but the room was never confirmed as reserved, so the room was rebooked.

Distributed transactions provide a way to spread the work across the appropriate components instead of trying to shoehorn everything into some omnibus component that duplicates functionality found elsewhere across the system.

# Comparing various methods of distributed transactions

In this section, we will look at three ways to handle consistency across a distributed system. The first will be the **Two-Phase Commit** (**2PC**), which can offer the strongest consistency but has some large drawbacks. The other two are the **Choreographed Saga** and the **Orchestrated Saga**, which still offer a good consistency model and are excellent options when 2PCs are not an option.

## The 2PC

At the center of a 2PC is a coordinator that sends the **Prepare** and **Commit** messages to all the participants. During the **Prepare** phase, each participant may respond positively to signify they have started a local transaction and are ready to proceed. If all the participants have responded positively, then the coordinator will send a COMMIT message to all of the participants and the distributed transaction will be complete. On the other hand, if any participant responds negatively during the **Prepare** phase, then the coordinator will send an ABORT message to inform the other participants to roll back their local transaction; again, the distributed transaction will be complete:

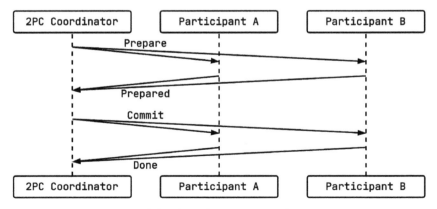

Figure 8.3 – A 2PC distributed transaction with two participants

What this method has going for it is that it is a widely known method and has a well-documented protocol for implementations to follow. When it is implemented correctly and used in a preferably very well-tested system, it can offer very strong consistency as it has all ACID guarantees.

During the **Prepare** phase, a participant would execute the following transaction in **PostgreSQL**:

```
BEGIN;
-- execute queries, updates, inserts, deletes ...
PREPARE TRANSACTION 'bfa1c57a-d99d-4d74-87a9-3aaabcc754ee';
```

Then, during the **Commit** phase, either a COMMIT or ABORT message would be received by each participant. Now, either a commit or a rollback of that prepared transaction would take place.

What the 2PC has going against it is big. During the **Prepare** phase, the participants all create prepared transactions that will consume resources until the coordinator gets around to sending the message for the **Commit** phase. If that never arrives for whatever reason, then the participants may end up holding open a transaction much longer than they should or may never resolve the transactions. Another possibility is that a participant may fail to properly commit the transaction, leaving the system in an inconsistent state. Holding onto transactions limits the scalability of this method for larger distributed transactions.

## The Saga

A saga is a sequence of steps that define the actions and compensating actions for the system components that are involved, also known as the saga participants. In contrast to 2PCs, each participant is not expected to use a prepared transaction. Not relying on prepared transactions opens the possibility of using NoSQL or another database that does not support prepared transactions. Sagas drop support for the isolation guarantee, making them ACD transactions. The saga steps may use a local ACID transaction, but any changes that are made will be visible to concurrent operations while the other steps are being run.

Another reason to choose a saga for your distributed transaction is that a saga can be long-lived. Since there are no resources tied up in a database blocking other work, we can build a saga that could have a lifetime of several seconds, minutes, or even longer:

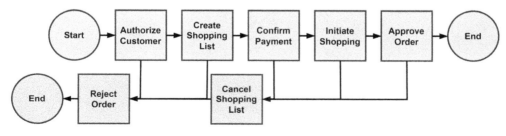

Figure 8.4 – A saga representing the create order process

In the preceding diagram, we have a saga representing the process of creating a new order. Along the top row are the actions we want to take to create the order. Along the bottom are the compensating actions that would be executed to roll back any changes to the system to bring it back to a consistent state.

A saga may be a collaborative effort between participants and be choreographed, alternatively there can be a saga execution coordinator that orchestrates the entire process.

Now, let's take a look at both types of sagas and how we might use either to handle creating new orders in the MallBots application.

### The Choreographed Saga

In a choreographed saga, each participant knows the role they play. With no coordinator to tell them what to do, each participant listens for the events that signal their turn. The coordination logic is spread out across the components and is not centralized.

Our example from *Figure 8.4* could be accomplished by publishing the following events into the message broker:

1.  The **Order Processing** module would publish an `OrderCreated` event after creating a new order in the pending state.

2.  The **Customers** module listens for `OrderCreated` events and publishes a `CustomerApproved` event after confirming the customer on the order.

3.  The **Depot** module also listens for the `OrderCreated` event and uses the order information to create a shopping list for the bots and publishes a `ShoppingListCreated` event.

4.  The **Payments** module listens for the `OrderCreated` and `CustomerApproved` events and will verify the authorized payment for the order and customer before publishing the `PaymentValidated` event.

5.  The **Depot** module will listen for the `PaymentValidated` event to hand the shopping list to a bot before publishing the `ShoppingInitiated` event.

6.  The **Order Processing** module will be listening for `ShoppingInitiated` to update the order state to "approved." Then, it will publish a final `OrderApproved` event.

These events and interactions, when mapped out, would look like this:

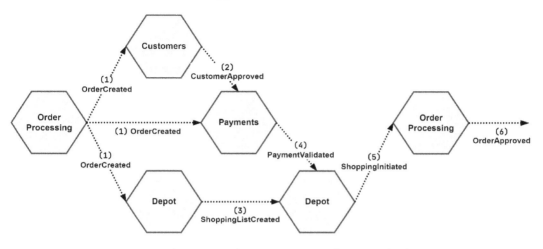

Figure 8.5 – The create order process using a choreographed saga

Compensation is initiated by participants listening to the events representing failures or other events representing undoing actions. If the attempt to validate the authorized payment for the order were to fail, then all the steps that modified any state would need to be rolled back.

1. The **Payments** module publishes an `UnauthorizedPayment` event after failing to validate the authorized payment with the information provided.

2. The **Depot** module is listening for the `UnauthorizedPayment` event and will cancel the shopping list before publishing the `ShoppingListCanceled` event.

3. The **Order Processing** module is also listening for the `UnauthorizedPayment` event and will reject the order, effectively canceling it in the process, before publishing the `OrderRejected` event.

The **Customers** module is not listening for the `UnauthorizedPayment` event because it has no way to react to the condition. It may also not be listening because the event was overlooked. Choreographed compensations can be tricky this way.

If the order approval, *(6)*, or shopping initiation, *(5)*, tasks were expected to fail, then the **Payments** module would need to listen for the events that would result from those failures so it could publish a compensation event. This way, the rest of the compensation would remain as-is. This would require coordination by the developers; miscommunication could be the source of saga failures.

Using a choreographed saga is a good choice when the number of participants is low, and the coordination logic is easy to follow. Choreography makes use of the events that participants already publish and subscribe to and does not need any extra services or processes to be deployed.

## The Orchestrated Saga

An orchestrated saga does not rely on individual components publishing events. Instead, it uses a **saga execution coordinator** (**SEC**) to send commands to the components. This centralizes the orchestration of the components into one location. When the coordinator receives a failed reply, it switches over to begin compensating and sending any compensation commands required to roll back the operation:

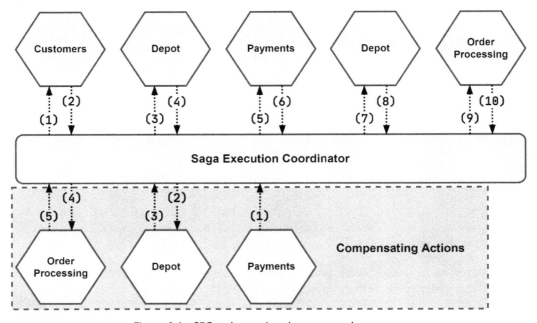

Figure 8.6 – SEC orchestrating the create order process

As shown in the preceding diagram, this operation would work with a saga orchestrated by an SEC, like so:

1.  The coordinator sends the `AuthorizeCustomer` command to the **Customers** module.
2.  The **Customers** module responds with a generic `Success` message.
3.  The coordinator sends the `CreateShoppingList` command to the **Depot** module.
4.  The **Depot** module responds with a `CreatedShoppingList` message.
5.  The coordinator sends the `ConfirmPayment` command to the **Payments** module.
6.  The **Payments** module responds with a generic `Success` message.
7.  The coordinator sends the `InitiateShopping` command to the **Depot** module.
8.  The **Depot** module responds with a generic `Success` message.

9.  The coordinator sends the `ApproveOrder` command to the **Order Processing** module.

10. The **Order Processing** module responds with a generic `Success` message.

The first time the SEC receives a response from the **Depot** module, the response is a specific **Depot** message – the `CreatedShoppingList` reply. This message contains the identity of the shopping list that was just created. The SEC adds that identity to the context of the saga so that it can be used later in the second call to **Depot** to initiate the shopping.

Handling compensation within an SEC is kicked off by any of the participants responding with a `Failure` message.

Starting again with a failure in the **Payments** module, the following must take place to compensate the saga:

1.  The **Payments** module would respond with a generic `Failure` message.

2.  The coordinator would begin the compensation process and send the `CancelShoppingList` command to the **Depot** module.

3.  The **Depot** module would respond with a generic **Success** message.

4.  The coordinator would send `RejectOrder` to the **Order Processing** module.

5.  The **Order Processing** module would respond with a generic `Success` message.

After the SEC receives the first `Failure` message, it expects each compensating action to complete successfully and responds with a `Success` message. The same would be true for the choreographed saga – each compensating action must complete without any issues.

The process of creating a new order is more than a couple of steps and more than one step is involved in compensating the transaction. So, in this case, it would be better to implement this process using an orchestrator than relying on choreography among the modules. To do that, we will need to add the supporting functionality to the application first.

## Implementing distributed transactions with Sagas

To organize the order creation process as a saga, we will be introducing additional functionality in the form of an SEC. These are the items we will be building or modifying to accomplish this task:

- We will update the `ddd` and `am` packages so that they include the new **Command** and **Reply** message types

- We will create a new `sec` package that will be the home for an orchestrator and saga definitions and implementations

Now, let's dive into the existing packages to add those new types of messages.

## Adding support for the Command and Reply messages

The **Command** and **Reply** additions to the ddd package are nearly exact copies of the **Event** definitions and implementations that we can expand on later. Here are the interfaces and implementations for **Reply**:

Figure 8.7 – The new Reply definitions in the ddd package

The ones for the **Command** message will be like the **Event** and **Reply** definitions shown in *Figure 8.7* with one small difference – the CommandHandler returns a **Reply**, along with an error:

```
CommandHandler[T Command] func(ctx context.Context, cmd T) (Reply, error)
```

Figure 8.8 – CommandHandler returns a Reply and an error

The additions to the am package are like the ones in the ddd package. The additions for CommandMessages will also be modified to return the replies, along with the errors:

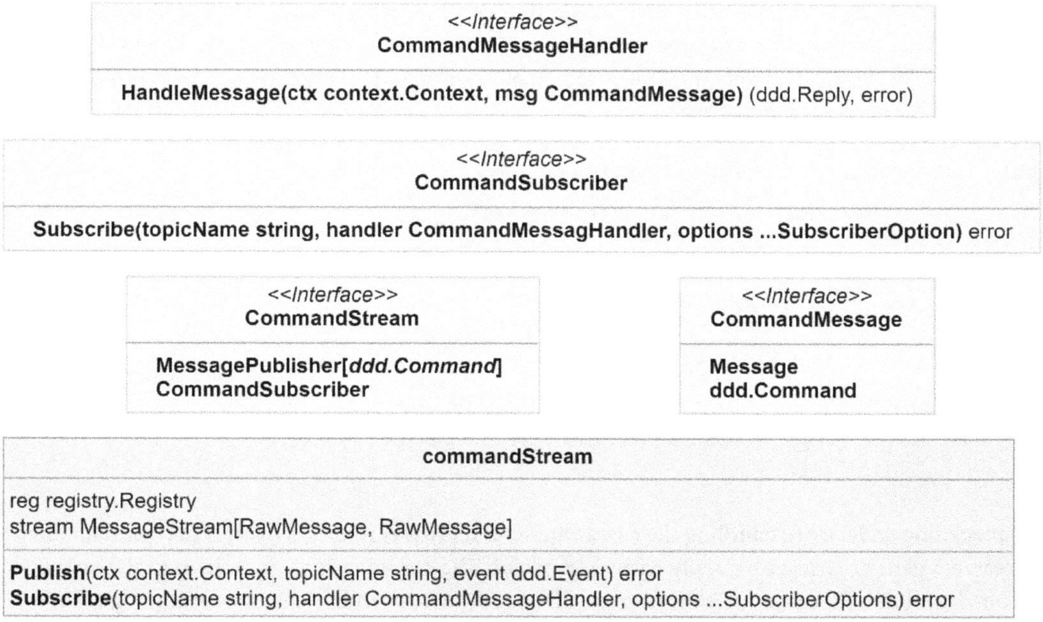

| <<*Interface*>> |
|---|
| **CommandMessageHandler** |
| **HandleMessage(ctx context.Context, msg CommandMessage)** (ddd.Reply, error) |

| <<*Interface*>> |
|---|
| **CommandSubscriber** |
| **Subscribe(topicName string, handler CommandMessagHandler, options ...SubscriberOption)** error |

| <<*Interface*>> | <<*Interface*>> |
|---|---|
| **CommandStream** | **CommandMessage** |
| **MessagePublisher[***ddd.Command***]**<br>**CommandSubscriber** | **Message**<br>**ddd.Command** |

| **commandStream** |
|---|
| reg registry.Registry<br>stream MessageStream[RawMessage, RawMessage] |
| **Publish**(ctx context.Context, topicName string, event ddd.Event) error<br>**Subscribe**(topicName string, handler CommandMessageHandler, options ...SubscriberOptions) error |

Figure 8.9 – The new CommandMessage definitions in the am package

When a **Command** message is handled, the expectation is that we will be responding with a **Reply** message. Instead of returning the result of handling **Command**, as we did in the EventStream implementation, we want to publish a reply. Before we do that, we need to determine if the overall outcome of **Command** was a success or a failure. We can determine that based on if an error was returned. Finally, the command handler might not have returned any reply, so a generic Success and Failure reply will be built and used in that case. This is how that is implemented:

```
reply, err = handler.HandleMessage(ctx, commandMsg)
if err != nil {
    return s.publishReply(ctx, destination,
        s.failure(reply, commandMsg),
    )
}
return s.publishReply(ctx, destination,
    s.success(reply, commandMsg),
)
```

A `CommandMessage` includes a special header that specifies where replies should be sent; that is how we get the `destination` value in the previous listing. Another special header is added to replies so that we can easily determine the outcome of a command. The following code shows how it can be added for successful outcomes:

```
func (s commandStream) success(
    reply ddd.Reply, cmd ddd.Command,
) ddd.Reply {
    if reply == nil {
        reply = ddd.NewReply(SuccessReply, nil)
    }
    reply.Metadata().Set(ReplyOutcomeHdr, OutcomeSuccess)
    return s.applyCorrelationHeaders(reply, cmd)
}
```

In the preceding code, we're handling the cases where no reply was returned by the command handler and created a generic `Success` reply with no payload. The commands we handle may also include other headers that help relate the action to specific aggregates, or in our case a running saga. So, we can also add those correlation headers from **Command** to **Reply**, as shown here:

```
func (s commandStream) applyCorrelationHeaders(
    reply ddd.Reply, cmd ddd.Command,
) ddd.Reply {
    for key, value := range cmd.Metadata() {
        if strings.HasPrefix(key, CommandHdrPrefix) {
            hdr := ReplyHdrPrefix + key[len
                (CommandHdrPrefix):]
            reply.Metadata().Set(hdr, value)
        }
    }
    return reply
}
```

New protocol buffer message declarations have also been added for `CommandMessageData` and `ReplyMessageData`. For simplicity, they are the same as the `EventMessageData` type.

That's all of the updates we need to make to the `am` package. Now, let's look at creating the new `sec` package.

# Adding an SEC package

The SEC is made of a few parts:

- An **Orchestrator**, which uses a saga for logic and connects it with a repository and publisher

- A **Saga Definition**, which holds all the metadata and the sequence of steps

- The steps that contain the logic for the actions, the **Reply handlers**, and their compensating counterparts:

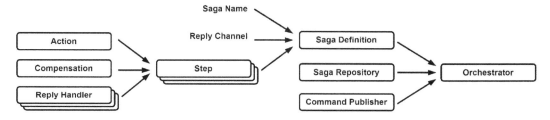

Figure 8.10 – How the SEC components come together

Much of this might seem novel or new if you are not very familiar with this pattern, so let's take a closer look at the three main parts of this implementation.

> **Using generics in the SEC**
>
> The implementations in the sec package make use of generics to allow the saga data payload to be used with ease in the actions and the **Reply** handlers that will need to be maintained in the application.

## *The Orchestrator*

The primary job of our **Orchestrator** implementation is to handle the incoming replies so that it can determine which step to execute, as well as when to fail over and begin compensating:

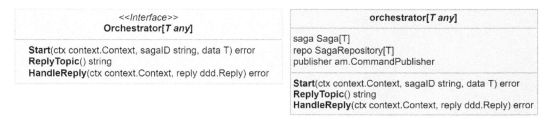

Figure 8.11 – The Orchestrator interface and struct definition

The Orchestrator has two modes of operation – a manual start and being reactive to the incoming replies that it receives. When a reply comes in, the outcome is looked at before which kind of action is executed on the current or next possible step is determined. After executing the action, if a **Command** is returned, then the Orchestrator will publish this **Command** to its destination.

## The Saga definition

The purpose of the **Saga** definition is to provide a single location for all of the logic on how the saga should operate:

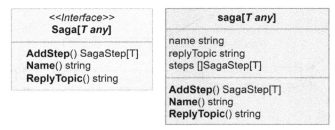

Figure 8.12 – The Saga interface and definition

Our **Saga** exists to hold the specifics and logic of the operation that needs to be distributed across the application. Each **Saga** that is running in the application will have a unique name and reply channel. Likewise, the sequence of steps will be unique to the saga definition, but the individual steps might not be.

## The Steps

**Steps** are where all the logic of a **Saga** is contained. They generate the **Command** messages that are sent to participants and can modify the data for the associated saga:

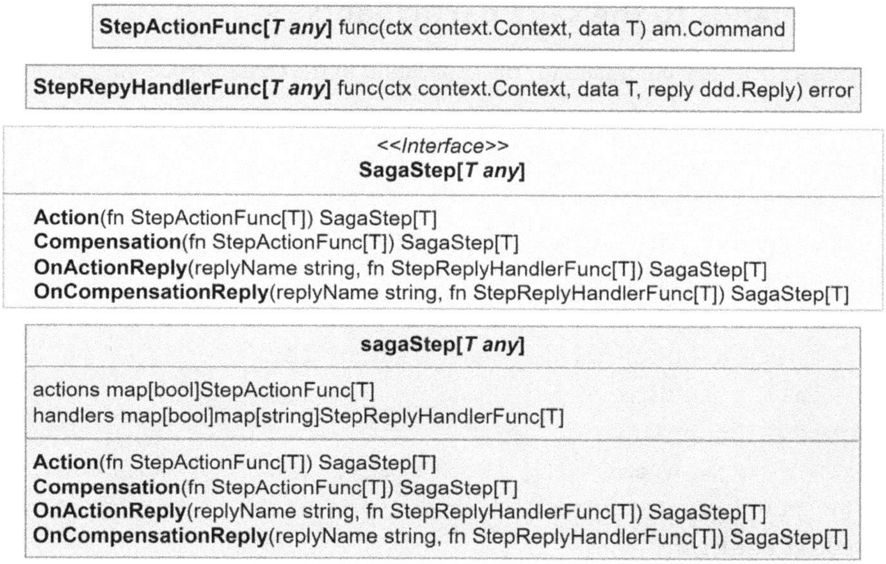

```
StepActionFunc[T any] func(ctx context.Context, data T) am.Command
```

```
StepRepyHandlerFunc[T any] func(ctx context.Context, data T, reply ddd.Reply) error
```

```
                        <<Interface>>
                        SagaStep[T any]

Action(fn StepActionFunc[T]) SagaStep[T]
Compensation(fn StepActionFunc[T]) SagaStep[T]
OnActionReply(replyName string, fn StepReplyHandlerFunc[T]) SagaStep[T]
OnCompensationReply(replyName string, fn StepReplyHandlerFunc[T]) SagaStep[T]
```

```
                        sagaStep[T any]

actions map[bool]StepActionFunc[T]
handlers map[bool]map[string]StepReplyHandlerFunc[T]

Action(fn StepActionFunc[T]) SagaStep[T]
Compensation(fn StepActionFunc[T]) SagaStep[T]
OnActionReply(replyName string, fn StepReplyHandlerFunc[T]) SagaStep[T]
OnCompensationReply(replyName string, fn StepReplyHandlerFunc[T]) SagaStep[T]
```

Figure 8.13 – The saga steps interface and definitions

Each **Step** has, at a minimum, either an action or compensating action defined, though a **Step** may also have both defined as well. Steps may add optional handlers for the replies that are being sent back by the participants to apply custom logic to them.

After the orchestrator is started or has processed a **Reply**, it will look for the next **Step** in the sequence that has defined an action for the given direction. Steps without any compensation actions will be skipped until either one is found or there are no more steps.

With that, we have the necessary functionality to convert the order creation process into an orchestrated saga. Let's take a look.

# Converting the order creation process to use a Saga

In this section, we will be implementing the create order process described earlier in this chapter as an orchestrated saga method. To do so, we will use the SEC from the previous section. We will be doing the following to accomplish this task:

- Updating the modules identified as participants to add new streams, handlers, and commands

- Creating a new module called `cosec`, short for *Create-Order-Saga-Execution-Coordinator*, that will be responsible for orchestrating the process of creating new orders

Let's begin by learning how to add commands.

## Adding commands to the saga participants

The existing `CreateOrder` command for the application in the **Order Processing** module looks like this:

```
order, err := h.orders.Load(ctx, cmd.ID)
// 1. authorizeCustomer
err = h.customers.Authorize(ctx, cmd.CustomerID)
if err != nil { return err }
// 2. validatePayment
err = h.payments.Confirm(ctx, cmd.PaymentID)
if err != nil { return err }
// 3. scheduleShopping
shoppingID, err = h.shopping.Create(ctx, cmd.ID, cmd.Items)
if err != nil { return err }
// 4. orderCreation
err = order.CreateOrder(
    cmd.ID, cmd.CustomerID, cmd.PaymentID,
    shoppingID, cmd.Items,
)
if err != nil { return err }
return h.orders.Save(ctx, order)
```

We will use the sequence of steps from the example for the orchestrated saga in the previous section to reimplement the previous listing as a **Saga**.

To create a new order, we need to create new **Command** messages in the following participants:

- The **Customers** module needs to implement the `AuthorizeCustomer` command
- The **Depot** module needs to implement the `CreateShoppingList`, `CancelShoppingList`, and `InitiateShopping` commands, as well as the reply from `CreatedShoppingList`
- The **Order Processing** module needs to implement the `ApproveOrder` and `RejectOrder` commands
- The **Payments** module needs to implement the `ConfirmPayment` command

Many of these commands have existing gRPC equivalents, so the easiest thing to do when we do implement them is to call the existing application instances we are creating in each module.

Let's look at each module and its share of commands in more detail.

## The Customers module

We have only one **Command** to define for the **Customers** module – a new protocol buffer message for `AuthorizeCustomer`:

```
message AuthorizeCustomer {
  string id = 1;
}
```

Just like the events we have already defined, we will create a constant called `AuthorizeCustomerCommand` that will hold the unique name for the command that will be used in its `Key()` method so that we can register the type in the registry. Alongside the string constant for the **Command** name, we also need to add a constant for the command channel for this module:

```
CommandChannel = "mallbots.customers.commands"
```

We will only need one **Command** channel for each module compared to the many possible channels that we created for the aggregates we published events to.

Our command handler will reside in the `commands.go` file in the `customers/internal/handlers` directory and will implement `ddd.CommandHandler[ddd.Command]`. It will also take advantage of the existing `application` command to authorize a customer:

| commandHandler |
| --- |
| app application.App |
| **HandleCommand**(ctx context.Context, cmd ddd.Command) (ddd.Reply, error) |

Figure 8.14 – The Customers module's command handler definition

We can handle the incoming commands in much the same way as we handled the event handlers in the previous chapters:

```
func (h commandHandlers) HandleCommand(
    ctx context.Context, cmd ddd.Command,
        ) (ddd.Reply, error) {
    switch cmd.CommandName() {
    case customerspb.AuthorizeCustomerCommand:
        return h.doAuthorizeCustomer(ctx, cmd)
    }
    return nil, nil
}
```

```
func (h commandHandlers) doAuthorizeCustomer(
    ctx context.Context, cmd ddd.Command,
) (ddd.Reply, error) {
    payload := cmd.Payload()
        .(*customerspb.AuthorizeCustomer)
    return nil, h.app.AuthorizeCustomer(
        ctx,
        application.AuthorizeCustomer{
            ID: payload.GetId(),
        },
    )
}
```

The doAuthorizeCustomer() method does not return any specific replies - only the generic Success and Failure ones. In the highlighted section of code, nil is being returned as the **Reply** value, and the result returned from AuthorizeCustomer() will be used to determine the outcome of handling the message. When that result is an error a Failure reply will be generated and returned.

In the same file, to make using the handlers easier, we can add a constructor and a function to register them:

```
func NewCommandHandlers(
    app application.App,
) ddd.CommandHandler[ddd.Command] {
        [ddd.Command] {
    return commandHandlers{
        app: app,
    }
}


func RegisterCommandHandlers(
    subscriber am.CommandSubscriber,
    handlers ddd.CommandHandler[ddd.Command],
) error {
    cmdMsgHandler := am.CommandMessageHandlerFunc(
        func(
            ctx context.Context,
```

```
        cmdMsg am.IncomingCommandMessage,
    ) (ddd.Reply, error) {
        return handlers.HandleCommand(ctx, cmdMsg)
    })
    return subscriber.Subscribe(
        customerspb.CommandChannel,
        cmdMsgHandler,
        am.MessageFilter{
            customerspb.AuthorizeCustomerCommand,
        },
        am.GroupName("customer-commands"),
    )
}
```

This command handler can be used for any commands and is not limited to only handling the commands coming from the create order saga. The **Customers** module remains uncoupled from the **Order Processing** module because we do not have any explicit ties to the **Order Processing** module in this handler. If we had other unrelated commands, we would also have them handled here in this command handler.

In the `module.go` file located at the root of the **Customers** module, we need to create a new **Command** stream, an instance of the **Command** handlers, and register the two together:

```
// setup Driven adapters
stream := jetstream.NewStream(mono.Config().Nats.Stream,
    mono.JS(), mono.Logger())
commandStream := am.NewCommandStream(reg, stream)
// setup application
commandHandlers := logging.LogCommandHandlerAccess
    [ddd.Command](
    handlers.NewCommandHandlers(app),
    "Commands", mono.Logger(),
)
// setup Driver adapters
err = handlers.RegisterCommandHandlers(
    commandStream, commandHandlers,
)
```

One module down, three to go! Thankfully, the work is going to be roughly the same for the remaining three:

- Define the commands, along with a constant containing the command channel for the module
- Create a command handler for the commands
- Wire up all the new things together in the composition root for the module

Now, let's look at the `Depot` module.

### The Depot module

The `Depot` module has three commands and a reply that we need to define. `CreateShoppingList` is a slightly interesting protocol buffer message:

```
message CreateShoppingList {
  message Item {
    string product_id = 1;
    string store_id = 2;
    int32 quantity = 3;
  }
  string order_id = 1;
  repeated Item items = 2;
}
```

What is interesting is that it is not a copy of the `OrderCreated` event from the **Order Processing** module. First, we do not have a `ShoppingId` that can be added yet. Second, we don't need to be generic and include requirements for data we don't need for a command in the `Depot` module. Something that's maybe not all that interesting but worth pointing out is that we did not copy and paste this message, forcing us to do unnecessary work.

The `CreateShoppingList` command when successfully handled will return a **Reply** with the identity of the newly created shopping list:

```
message CreatedShoppingList {
  string id = 1;
}
```

Since this **Command** returns a specific **Reply**, this means we do not handle it as we did in the **Customers** module for `AuthorizeCustomer`:

```
func (h commandHandlers) doCreateShoppingList(
    ctx context.Context, cmd ddd.Command,
```

```
        ) (ddd.Reply, error) {
    payload := cmd.Payload().(*depotpb.CreateShoppingList)
    id := uuid.New().String()
    // snip build items ...
    err := h.app.CreateShoppingList(
        ctx,
        commands.CreateShoppingList{
            ID:      id,
            OrderID: payload.GetOrderId(),
            Items:   items,
        },
    )
    return ddd.NewReply(
        depotpb.CreatedShoppingListReply,
        &depotpb.CreatedShoppingList{Id: id},
    ), err
}
```

This time, we return `CreatedShoppingListReply` and a possible error. Admittingly, this is another shortcut, but if there was an error, then the **Reply** message we send will not be handled unless there is also a handler for it that was added to the compensating side.

### The Order Processing module

The two commands that we are using in the **Order Processing** module do not have existing gRPC or application command implementations, so we will need to add them to the application.

For the `ApproveOrder` command, we will be receiving `ShoppingID` from the `Depot` module, which it sends back in its `CreatedShoppingList` **Reply** message. For `RejectOrder`, the content is simply the identity of the order that was being created that now needs to be rejected.

If you have forgotten how we implemented application commands for the **Order Processing** module, here is a quick refresher by way of the `ApproveOrder` command in the `ordering/internal/application/commands` directory:

```
type ApproveOrder struct {
    ID         string
    ShoppingID string
}
type ApproveOrderHandler struct {
```

```
        orders   domain.OrderRepository
    publisher ddd.EventPublisher[ddd.Event]
}

func NewApproveOrderHandler(
    orders domain.OrderRepository,
    publisher ddd.EventPublisher[ddd.Event],
) ApproveOrderHandler {
    return ApproveOrderHandler{
        orders:   orders,
        publisher: publisher,
    }
}
func (h ApproveOrderHandler) ApproveOrder(
    ctx context.Context, cmd ApproveOrder,
) error {
    order, err := h.orders.Load(ctx, cmd.ID)
    event, err := order.Approve(cmd.ShoppingID)
    err = h.orders.Save(ctx, order)
    return h.publisher.Publish(ctx, event)
}
```

This command handler is plugged into the `application.Commands` interface and the `application.appCommands` struct with initialization in the `Application` constructor. This makes it available to the command message handler, as well as to the gRPC server if we decide to add it there as well.

We can add the parts that handle the command messages by completing the same steps we did for the last two modules. Here, we must define the commands and handlers and update the composition root to bring it all together.

### The Payments module

There is nothing noteworthy about adding command message handlers to the **Payments** module since I covered all the unusual cases in the previous three module sections.

I will close out this section with a checklist for adding command handlers to a module:

- Add **Command** and **Reply** protocol buffer message declarations
- Create name constants for each **Command** and **Reply**

- Create the `Key()` methods for each **Command** and **Reply**

- Include each **Command** and **Reply** in the `Registrations()` function

- Create new application commands if they are not already implemented

- Create a command message handler and handle each command

- Update the composition root to create a command stream

- Update the composition root to create an instance of the command handlers

- Update the composition root to register the handlers with the stream

After the **Payments** modules has been updated to handle commands, we are ready to orchestrate the modules together to create our orders.

## Implementing the create order saga execution coordinator

Creating an order in the **Order Processing** module is triggered by the `BasketCheckOut` event. We can continue to do that in **Order Processing**. In this section, we will be implementing the saga in a new module called `cosec` that will be reactive to the `OrderCreated` event from the **Order Processing** module.

> **Why not trigger the Saga off the BasketCheckedOut event?**
>
> We could have and it would work mostly the same with maybe an additional step or alternate action or two. I will leave reimplementing the Saga that way as an exercise for you.

### Registering all the external types

The saga will be sending commands and receiving replies from a handful of modules. So, in the composition root in the **Driven** adapters section, after the registry has been created, we have the following:

```
err = orderingpb.Registrations(reg)
if err != nil { return }
err = customerspb.Registrations(reg)
if err != nil { return }
err = depotpb.Registrations(reg)
if err != nil { return }
err = paymentspb.Registrations(reg)
if err != nil { return }
```

Each module that participates in the saga can be seen in the preceding code. This makes all the commands and replies available to use. If we were implementing a choreographed saga, then these external module registrations would need to be included in all the correct places, which could potentially be a bit of a maintenance nightmare.

### Defining the saga data model

Our saga will need to keep track of the order and other related facts:

```
type CreateOrderData struct {
    OrderID    string
    CustomerID string
    PaymentID  string
    ShoppingID string
    Items      []Item
    Total      float64
}

type Item struct {
    ProductID string
    StoreID   string
    Price     float64
    Quantity  int
}
```

The `CreateOrderData` struct will be used in all those places where generics were used in the `sec` package.

### Adding the saga repository

The saga repository works a little like `AggregateRepository`, where there is an infrastructure-specific store implementation that we will use to read and write the data:

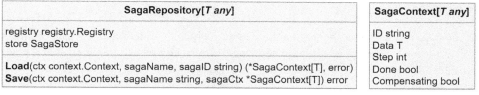

| SagaRepository[*T any*] |
| --- |
| registry registry.Registry<br>store SagaStore |
| **Load**(ctx context.Context, sagaName, sagaID string) (*SagaContext[T], error)<br>**Save**(ctx context.Context, sagaName string, sagaCtx *SagaContext[T]) error |

| SagaContext[*T any*] |
| --- |
| ID string<br>Data T<br>Step int<br>Done bool<br>Compensating bool |

| <<*Interface*>><br>**SagaStore** |
| --- |
| **Load**(ctx context.Context, sagaName, sagaID string) (*SagaContext[[]byte], error)<br>**Save**(ctx context.Context, sagaName string, sagaCtx *SagaContext[[]byte]) error |

Figure 8.15 – The saga repository definition, store interface, and context model

For PostgreSQL, we are using the following table schema:

```
CREATE TABLE cosec.sagas (
 id            text    NOT NULL,
 name          text    NOT NULL,
 data          bytea   NOT NULL,
 step          int     NOT NULL,
 done          bool    NOT NULL,
 compensating  bool    NOT NULL,
 updated_at    timestamptz NOT NULL DEFAULT
  CURRENT_TIMESTAMP, PRIMARY KEY (id, name)
);
```

We must use the following few lines to create the store and repository in our composition root:

```
sagaStore := pg.NewSagaStore("cosec.sagas", mono.DB(), reg)
sagaRepo := sec.NewSagaRepository[*models.CreateOrderData](
    reg, sagaStore,
)
```

The saga data generic is defined with a pointer so that it can be modified by the functions we will be adding to the saga steps.

## Defining the saga

To define the saga, we need to set the saga name, the saga reply channel, and the steps that are involved with the operation we want to run. Without showing the steps and related methods, this is how the saga is created:

```
const CreateOrderSagaName     = "cosec.CreateOrder"
const CreateOrderReplyChannel = "mallbots.cosec.replies"
type createOrderSaga struct {
    sec.Saga[*models.CreateOrderData]
}

func NewCreateOrderSaga() sec.Saga[*models.CreateOrderData] {
    saga := createOrderSaga{
        Saga: sec.NewSaga[*models.CreateOrderData](
            CreateOrderSagaName,
            CreateOrderReplyChannel,
        ),
    }
    // steps go here
    return saga
}
```

We can define the saga using the Builder pattern (`https://refactoring.guru/design-patterns/builder/go/example`). This would look something like this in an extreme case:

```
saga.AddStep().
    Action(actionCommandFn).
    ActionReply("some.Reply", onSomeReplyFn).
    ActionReply("other.Reply", onOtherReplyFn).
    Compensation(compensationCommandFn).
    CompensationReply("nope.Reply", onNopeReply)
```

The preceding example demonstrates each possible modification that can be made to a step. Steps may have many reply handlers:

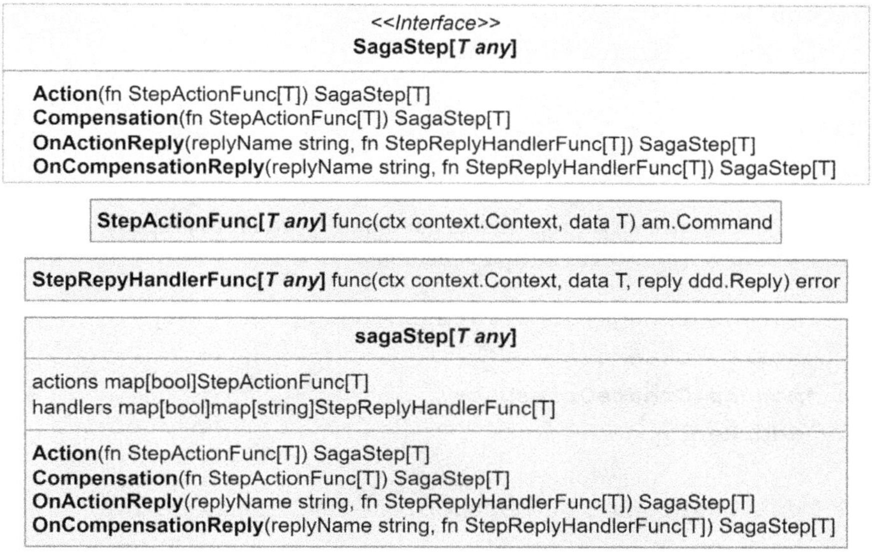

Figure 8.16 – The SagaStep interface and related types

For the create order saga, the following steps must be defined:

```
// 0. -RejectOrder
saga.AddStep().
    Compensation(saga.rejectOrder)
// 1. AuthorizeCustomer
saga.AddStep().
    Action(saga.authorizeCustomer)
// 2. CreateShoppingList, -CancelShoppingList
saga.AddStep().
    Action(saga.createShoppingList).
    OnActionReply(
        depotpb.CreatedShoppingListReply,
        saga.onCreatedShoppingListReply,
    ).
    Compensation(saga.cancelShoppingList)
// 3. ConfirmPayment
saga.AddStep().
    Action(saga.confirmPayment)
// 4. InitiateShopping
```

```
saga.AddStep().
    Action(saga.initiateShopping)
// 5. ApproveOrder
saga.AddStep().
    Action(saga.approveOrder)
```

We can use the methods defined on the saga so that our data is typed for us properly, as shown in the following onCreatedShoppingListReply method:

```
func (s createOrderSaga) onCreatedShoppingListReply(
    ctx context.Context,
    data *models.CreateOrderData,
    reply ddd.Reply,
) error {
    p := reply.Payload().(*depotpb.CreatedShoppingList)
    data.ShoppingID = p.GetId()
    return nil
}
```

The **Reply** payloads will still need to be cast to the correct types before you can work with them.

The methods provided to Action() and Compensation() generate the commands that our saga participants must carry out for us. For an example of an action, we can look at the confirmPayment() method that is used to generate and send the command to confirm that the payment was authorized properly:

```
func (s createOrderSaga) confirmPayment(
    ctx context.Context, data *models.CreateOrderData,
) am.Command {
    return am.NewCommand(
        paymentspb.ConfirmPaymentCommand, // command name
        paymentspb.CommandChannel, // command destination
        &paymentspb.ConfirmPayment{ // command payload
            Id:     data.PaymentID,
            Amount: data.Total,
        },
    )
}
```

The `ConfirmPaymentCommand` command in the preceding code is intended for the **Payments** module. It is defined in that module because our orchestrator does not own the commands that it publishes. A module's commands and replies should be documented alongside its published and subscribed events.

## Creating the message handlers

As I said at the start of this section, we will be listening for the `OrderCreated` integration event from the **Order Processing** module as our trigger. The design of the actual handler itself is just like the others, but we start a saga instead of creating a data cache or executing some application command:

```
func (h integrationHandlers[T]) onOrderCreated(
    ctx context.Context, event ddd.Event,
) error {
    payload := event.Payload().(*orderingpb.OrderCreated)
    // compute items and total
    data := &models.CreateOrderData{
        OrderID:    payload.GetId(),
        CustomerID: payload.GetCustomerId(),
        PaymentID:  payload.GetPaymentId(),
        Items:      items,
        Total:      total,
    }
    return h.orchestrator.Start(ctx, event.ID(), data)
}
```

The preceding code starts the saga, which as its first action will locate the next step it should execute and then run it. The orchestrator does not keep the sagas running or in memory. After each interaction, the orchestrator will write the saga context into the database and return it to the reply message handler; as I mentioned earlier, they are reactive.

This is made clear with the following code, which registers the orchestrator as a reply message handler:

```
func RegisterReplyHandlers(
    subscriber am.ReplySubscriber,
    o sec.Orchestrator[*models.CreateOrderData],
) error {
    h := am.MessageHandlerFunc[am.IncomingReplyMessage](
        func(
            ctx context.Context,
```

```
            replyMsg am.IncomingReplyMessage,
        ) error {
            return o.HandleReply(ctx, replyMsg)
        },
    )
    return subscriber.Subscribe(
        o.ReplyTopic(),
        h,
        am.GroupName("cosec-replies"),
    )
}
```

The orchestrator handler replies directly, so it is not necessary to create another reply handler intermediary in the module's composition root. We only need a handler for the reply message that calls the orchestrator to handle the reply.

### Updating the composition root

Back in the composition root, we need to create streams for all the messages we intend to receive, including events, commands, and replies:

```
stream := jetstream.NewStream(
    mono.Config().Nats.Stream, mono.JS(), mono.Logger(),
)
eventStream := am.NewEventStream(reg, stream)
commandStream := am.NewCommandStream(reg, stream)
replyStream := am.NewReplyStream(reg, stream)
```

This module will not have an application like the rest of the existing modules, but we must still create the necessary handlers:

```
orchestrator := logging.LogReplyHandlerAccess
    [*models.CreateOrderData](
    sec.NewOrchestrator[*models.CreateOrderData](
        internal.NewCreateOrderSaga(),
        sagaRepo,
        commandStream,
    ),
    "CreateOrderSaga", mono.Logger(),
```

```
)
integrationEventHandlers := logging.LogEventHandlerAccess
    [ddd.Event](
    handlers.NewIntegrationEventHandlers(orchestrator),
    "IntegrationEvents", mono.Logger(),
)
```

Now, these handlers need to be wired up with the streams that will be driving them:

```
err = handlers.RegisterIntegrationEventHandlers(
    eventStream, integrationEventHandlers,
)
if err != nil { return err }
err = handlers.RegisterReplyHandlers(
    replyStream, orchestrator,
)
if err != nil { return err }
```

That's it – we have a working orchestrator and saga that will coordinate the creation of new orders. Here, we created a new module that will have an orchestrator running that will take care of the distributed operation that creates new orders.

This orchestrator did not need to have its own module – it would work just the same had it been built inside the **Order Processing** module. I simply chose to implement it in a module on its own so that the demonstration was clearer, and so that no details got lost in the other details of the existing module.

The existing CreateOrderHandler.CreateOrder() method in the **Order Processing** module still needs to be updated. When executing its tasks, it should no longer make any calls to external systems. This can be seen in the following code with error handling removed:

```
func (h CreateOrderHandler) CreateOrder(ctx
    context.Context, cmd CreateOrder) error {
    order, _ := h.orders.Load(ctx, cmd.ID)
    event, _ := order.CreateOrder(
        cmd.ID, cmd.CustomerID, cmd.PaymentID, cmd.Items,
    )
    _ = h.orders.Save(ctx, order)
    return h.publisher.Publish(ctx, event)
}
```

Now, without any calls to external services, creating an order is much more resilient. The saga is also more resilient, which means it will not be bothered by a service being down; so long as it eventually comes back up, the saga will also eventually get back to executing the steps it needs to.

## Summary

In this chapter, we learned about the challenges that you may face when working with distributed systems and dealing with work or operations that cannot be accomplished by a single component and must also be distributed. We looked at three methods and how their distributed workflows can be implemented – 2PCs, choreographed sagas, and orchestrated sagas. Finally, we implemented the existing create order operation using an orchestrated saga, which resulted in a more resilient process.

In the next chapter, we will learn how to improve resiliency for the entire system. To do so, we will learn about the different transactional boundaries that exist in distributed systems and more.

# 9

# Transactional Messaging

In this book, we have transformed an application into an asynchronous one, which removes a lot of issues that arose from the tightly coupled and temporally bound nature of synchronous communication. Nearly all communication in the application is now made with a message brokered through NATS JetStream, providing loose coupling for the application components. However, despite all the advances we have made, we still face issues that all distributed systems suffer from.

In this chapter, we are going to discuss the following main topics:

- Identifying problems faced by distributed applications
- Exploring transactional boundaries
- Using an Inbox and Outbox for messages

## Technical requirements

You will need to install or have installed the following software to run the application or to try the examples:

- The Go programming language, version 1.18+
- Docker

The code for this chapter can be found at `https://github.com/PacktPublishing/Event-Driven-Architecture-in-Golang/tree/main/Chapter09`.

## Identifying problems faced by distributed applications

In every distributed application there are going to be places where interactions take place between components that reside in different bounded contexts or domains. These interactions come in many forms and can be synchronous or asynchronous. A distributed application could not function without a method of communication existing between the components.

In the previous chapter, we looked at some different ways, such as using sagas, to improve the overall reliability of complex operations that involve multiple components. The reliability we added is at the operation level and spans multiple components, but it does not address reliability issues that happen at the component level.

Let's take a look at what affects reliability in synchronous and asynchronous distributed applications and what can be done to address the problem.

## Identifying problems in synchronous applications

In the first version of the MallBots application from much earlier in the book, we only had synchronous communication between the components, and that looked something like this:

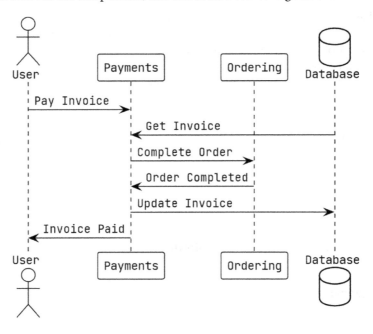

Figure 9.1 – Synchronous interaction between the Payments and Ordering modules

In the preceding example, the **Payments** module receives a request for an invoice to be marked as paid. When an invoice is to be paid, a synchronous call is made to the **Ordering** module to also update the order status to Completed. Before returning to the user, the updated invoice is saved back into the database.

## Identifying problems in asynchronous applications

After fully updating the application to use asynchronous communication, we see that the same action now looks like this:

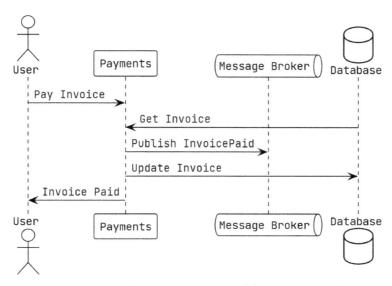

Figure 9.2 – Asynchronous interactions of the Payment module

In the updated version of the `Pay Invoice` action, we publish the `InvoicePaid` event instead of calling the **Ordering** module directly. The **Ordering** module is listening for this event, and when it receives the event, it will take the same action as before.

In both implementations, we can run into problems if the updated invoice data is not saved in the database. Also, for both implementations, you might think that a solution such as making a second call to the **Ordering** module or publishing an event to revert the change might help. However, there are a number of things that would make that solution improbable:

- The `Ordering` module has made or triggered irreversible modifications

- The second call or published event could also fail

- The failure is exceptional and there is no opportunity to take any corrective action

Adopting an asynchronous event-driven approach has not necessarily made the situation any better. In fact, things might be worse off now. When we publish the `InvoicePaid` event, we no longer directly know who has consumed the event and which other changes have happened as a result.

This is called a dual write problem, and it will exist everywhere we make a change in the application state that requires the change to be split in two or more places. When a write fails while multiple writes are being made, the application can be left in an inconsistent state that can be difficult to detect.

## Examining potential ways to address the problem

Sagas provide distributed transaction-like support for applications, and when an error occurs, they can be compensated. Using a small two-step saga might help here, but sagas are an operation consistency solution and not a component consistency solution. They cannot help with recovering a missing message or missing write in the database. That would be like using a sledgehammer to swat a fly.

Reordering the writes so that the more likely-to-fail write happens first does leave the more reliable ones to follow, but swapping the order of the writes will not help because the second write to whichever destination it is can always fail. Unless the action that is reordered to happen first is of little to no consequence, reordering the actions is not going to really be of any help.

Writing everything into the database while using a transaction could help if everything we were writing was in the database. However, in both of our examples, we are dealing with either a gRPC call or the publishing of a message. If we could convert all of our writes into ones that could be directed at our database, this approach would hold some promise.

## The singular write solution

If writing into multiple destinations is the problem, then it is reasonable to think that writing to a single destination could be a solution. Even more important than having a single write is that we need to have a single transactional boundary around what we will be writing.

We need a mechanism to make sure that the messages we publish must be added to the database in addition to or instead of being sent to NATS JetStream. We also need a way to create a single transaction boundary for this write to take place in. This means that for our application, we must create a transaction in PostgreSQL, into which we will put our writes in order to combine them into a single write operation.

This is called the *Transactional Outbox pattern*, or sometimes just the *Outbox pattern*. With it, we will be writing the messages we publish into a database alongside all of our other changes. We will be using a transaction so that the existing database changes that we would normally be making, and the messages, are written atomically. To do this, we will need to make the following changes to the modules:

- Setting up a single database connection and transaction that is used for the lifetime of an entire request
- Intercepting and redirecting all outgoing messages into the database

With these two items, we have our plan. First up, will be looking at the implementation of a transactional boundary around each request, followed by the implementation of a transactional outbox.

# Exploring transactional boundaries

Starting with the more important part first, we will tackle how to create a new transaction for each request into our modules, whether these come in as messages, a gRPC call, or the handling of a domain event side effect. As we are using `grpc-gateway`, all of the HTTP requests are proxied to our gRPC server and will not need any special attention.

Creating a transaction is not the difficult part. The challenge will be ensuring the same transaction is used for every database interaction for the entire life of the request. With Go, our best option is going to involve using the context to propagate the transaction through the request. Before going into what that option might look like, we should also have a look at some of the other possible solutions:

- We can toss out the option of using a global variable right away. Beyond being a nightmare to test, they will also become a problem to maintain or refactor as the application evolves.

- A new parameter could be added to every method and function to pass the transaction along so that it can eventually be passed into the repositories and stores. This option, in my opinion, is completely out of the question because the application would become coupled to the database. It also would go against the clean architecture design we are using and make future updates more difficult.

- A less ideal way to use a context with a transaction value within it would be to modify each of our database implementations to look for a transaction in the provided context. This would require us to update every repository or store and require all new implementations to look for a transaction. Another potential problem with this is we cannot drop in any third-party database code because it will not know to look for our transaction value.

A more ideal way to use the context, in my opinion, is to create a repository and store instances when we need them and to pass in the transaction using a new interface in place of the `*sql.DB` struct we are using now. Using a new interface will be easy and will result in very minimal changes to the affected implementations. Getting the repository instances created with the transactions will be handled by a new **dependency injection** (**DI**) package that we will be adding. The approach I am going to take will require a couple of minor type changes in the application code with the rest of the changes all made to the composition roots and entry points.

## How the implementation will work

A new DI package will be created so that we can create either singleton instances for the lifetime of the application or scoped instances that will exist only for the lifetime of a request. We will be using a number of scoped instances for each request so that the transactions we use can be isolated from other requests.

Before jumping further into how this implementation works, I should mention that this approach is not without a couple of downsides. The first is from the jump in complexity that using a DI package brings to the table. We will also be making use of a lot of type assertions because the container will be unaware of the actual types it will be asked to deliver.

Most of the updates will be made to the composition roots, which I believe softens the impact of those downsides somewhat.

Our implementation is going to require a new DI package. We will want the package to provide some basic features such as the management of scoped resources so that we can create our transactional boundaries.

## The di package

The `internal/di` package will provide a container that will be used to register factory functions for singletons and scoped values, as illustrated in the following screenshot:

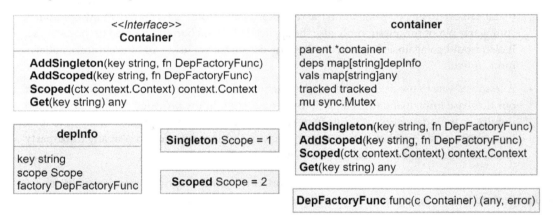

Figure 9.3 – The container type, interface, and dependencies

At the center of the `di` package is the container, which is accessed everywhere using the exported interface of the same name. In the composition roots, we will initialize a new container and then use the `AddSingleton()` and `AddScoped()` methods with factories for every dependency we will require.

In the handler code, we will be using the `Scoped()` function, which takes in a context and returns a new context with the container added as a value. It will be these contexts that will enable us to call upon resources created on a per-request basis.

There is also a `di.Get()` function that is used to fetch values from the container inside the contexts:

```
Get(ctx context.context, key string) any
```

The values returned by either Get() will need to be typecast before they are used. Both the di.Get() function and the Get() method on the container will panic if the key provided has not been registered. The reason the two will panic is that this is essentially the same as the startup code, which should halt the application to keep it from running in an invalid state.

Now that we are aware of what the di package will offer, we can learn more about the purposes of each part, starting with the containers.

## Setting up a container

To set up a new container, we use the di.New() function to create one and then use either AddSingleton() or AddScoped() to add our dependencies:

```
container := di.New()
container.AddSingleton("db",
    func(c di.Container) (any, error) {
        return mono.DB(), nil
    },
)
container.AddScoped("tx",
    func(c di.Container) (any, error) {
        db := c.Get("db").(*sql.DB)
        return db.Begin()
    },
)
```

I am choosing to use the short type assertion syntax when I use Get() here. I skip the type assertion checks since they are used so often that simple test runs would reveal problems if the wrong types were used.

## Setting the lifetime of dependencies

To register a dependency, you will use either AddSingleton() or AddScoped() with a string for the key or dependency name, and a factory function that returns either the dependency or an error. You can continue to build dependencies as a graph using the container value that is passed in. Here's an example of building a repository instance using the database:

```
repo := container.AddScoped("some-repo",
    func(c di.Container) (any, error) {
        db := c.Get("db").(*sql.DB)
        return postgres.NewSomeRepo(db), nil
```

```
        },
    )
```

Dependencies are grouped into two buckets:

- Singleton instances that are created once for the lifetime of the application. The same instance will be provided to each call for that dependency.

- Scoped instances that will be recreated the first time they are called up for each new scope. When the scope is no longer needed, the scoped instances will be available for garbage collection.

Here is a look at how the singleton and scoped instances along with the container will be used in the composition root after we are done transforming it:

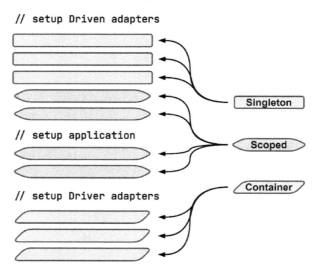

Figure 9.4 – Dependency and container usage in the composition root

When we are done updating the composition root, the driven adapters section will be a mix of singletons and scoped dependencies. The application section will be entirely made up of scoped dependencies, and the driver adapters section will be updated to use the container instead of any dependencies directly.

## *Using scoped containers*

We will use the `Scoped()` method to create a new child container that is then added as a value to the context provided to `Scoped()`. The current container is added to the child container as its parent container. The parent container will be used to locate singleton dependencies. The returned context should be passed into everything to propagate the container throughout the request. The code is illustrated in the following snippet:

```
container := di.New()

container.AddSingleton("db", dbFn)
container.AddScoped("tx", txFn)

db1 := container.Get("db")
tx1 := container.Get("tx")

ctx := container.Scoped(context.Background())

db2 := di.Get(ctx, "db") // same instance as db1
tx2 := di.Get(ctx, "tx") // entirely new instance
```

Dependencies that we have declared as singletons will always return the same instance for every call. Scoped dependencies will return a new instance for each scope they are needed in. A second call for a scoped dependency—for example, the `tx` dependency, from the same scoped container—will return the same instance as the first call.

Next, we will dive into a module to switch its composition root over to using the DI containers.

## Updating the Depot module with dependency containers

Using the `di` package in each of the modules is going to be the same but for the examples of what needs to be updated, I am going to use the **Depot** module. I have chosen the **Depot** module because it uses every kind of event and message handler.

We first create a new container at the start of the composition root `Startup()` method, like so:

```
func (Module) Startup(
    ctx context.Context, mono monolith.Monolith
) (err error) {
    container := di.New()
    // ...
}
```

In the composition root, we will tackle the changes in three parts. First, the driven adapters need to be divided up into singleton and scoped dependencies, then the application and handlers need to be also made into dependencies, and finally, we will update the servers or handler registration functions to use the container to create new instances of the application as needed for each request. Let us explore this further.

### Driven adapters

The factories we use for dependencies, such as the registry, will include the initialization code so that we continue to only execute it the one time, as shown here:

```
container.AddSingleton("registry",
    func(c di.Container) (any, error) {
        reg := registry.New()
        err := storespb.Registrations(reg)
        if err != nil { return nil, err }
        err = depotpb.Registrations(reg)
        if err != nil { return nil, err }
        return reg, nil
    },
)
```

We can go down the line of adapters, turning each one into a singleton dependency until we reach the point where we need to create a shoppingLists dependency:

```
shoppingLists := postgres.NewShoppingListRepository(
    "depot.shopping_lists",
    mono.DB(),
)
```

We want to use a transaction for this table and for all the others, but we cannot simply replace the database connection with a transaction. The following factory would not work out how we'd expect it to:

```
container.AddScoped("shoppingLists",
    func(c di.Container) (any, error) {
        return postgres.NewShoppingListRepository(
            "depot.shopping_lists",
            mono.DB().Begin(),
        ), nil
    },
)
```

Granted, the prior listing would create a new transaction every time we created this dependency for the scope. However, only the shoppingLists dependency would be using the transaction. All other database interactions would not be part of that transaction. We need to instead define a new scoped dependency for the transaction itself:

```
container.AddScoped("tx",
    func(c di.Container) (any, error) {
        return mono.DB().Begin()
    },
)
```

The tx dependency can now be injected into the dependencies that need a database connection. This switch to using transactions is what necessitates a small field type change in the application code. In all of the repository implementations, we have used the *sql.DB type for the db fields and we want to now pass in a *sql.Tx type.

To allow this, a new interface is added to the shared internal/postgres package that can be used to allow either a *sql.DB or *sql.Tx type to be used, as illustrated here:

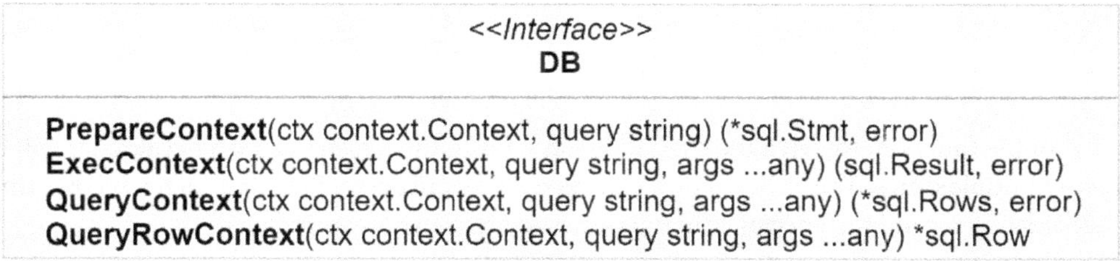

| <<*Interface*>> |
| --- |
| **DB** |
| **PrepareContext**(ctx context.Context, query string) (*sql.Stmt, error)<br>**ExecContext**(ctx context.Context, query string, args ...any) (sql.Result, error)<br>**QueryContext**(ctx context.Context, query string, args ...any) (*sql.Rows, error)<br>**QueryRowContext**(ctx context.Context, query string, args ...any) *sql.Row |

Figure 9.5 – The new DB interface that replaces *sql.DB and *sql.Tx

The new DB interface can be used to replace every usage of *sql.DB in our repository and store implementations so that then we can use either a database connection or a transaction. We can now correctly create a shoppingLists dependency:

```
container.AddScoped("shoppingLists",
    func(c di.Container) (any, error) {
        return postgres.NewShoppingListRepository(
            "depot.shopping_lists",
            c.Get("tx").(*sql.Tx),
        ), nil
    },
)
```

Stores such as `EventStore` and `SnapshotStore` are also updated to use the new DB interface in place of the `*sql.DB` type—for example, from the DI updates made to the **Ordering** module:

```
container.AddScoped("aggregateStore",
    func(c di.Container) (any, error) {
        tx  := c.Get("tx").(*sql.Tx)
        reg := c.Get("registry").(registry.Registry)
        return es.AggregateStoreWithMiddleware(
            pg.NewEventStore(
                "ordering.events",
                tx, reg,
            ),
            pg.NewSnapshotStore(
                "ordering.snapshots",
                tx, reg,
            ),
        ), nil
    },
)
```

At the start of the function, we fetch the transaction as `tx` and the registry as `reg` because we will be using them multiple times. This is being done for readability purposes. Fetching each a second time would not create a second instance of either dependency or cause any problems.

### Application and handlers

There will be no surprises in turning the application and each handler into a scoped dependency. For example, for `CommandHandlers`, we do the following:

```
container.AddScoped(
    "commandHandlers",
    func(c di.Container) (any, error) {
        return logging.
            LogCommandHandlerAccess[ddd.Command](
            handlers.NewCommandHandlers(
                c.Get("app").(application.App),
            ),
            "Commands",
            c.Get("logger").(zerolog.Logger),
```

```
        ), nil
    },
)
```

The application and each handler will need to be specified as a scoped dependency because we need to be able to create new instances for every request the module receives.

That leaves the driver adapters as the last part of the composition root that needs to be updated.

## Driver adapters

The driver adapters had been using the various variables we had created in the first two sections, but those variables no longer exist. Every driver needs to be updated to accept the container instead.

We will leave the existing driver functions alone and will create new functions that take the container instead. For example, the `RegisterDomainEventHandlers()` function will be replaced with a new function with the following signature:

```
func RegisterDomainEventHandlersTx(container di.Container)
```

The gRPC server and the three handlers will each need to be updated to make use of the container and to start a new scoped request.

### Updating the gRPC server

The new function to register our gRPC server will have the following signature, which swaps out the `application.App` parameter for the container:

```
func Register(
    container di.Container,
    registrar grpc.ServiceRegistrar,
) error
```

This function will create a new server called `serverTx` that is built with the container instead of the application instance. Like the existing server, it will implement `depotpb.DepotServiceServer`, but it will proxy all calls into instances of the server that are created for each request:

```
func (s serverTx) CreateShoppingList(
    ctx context.Context,
    request *depotpb.CreateShoppingListRequest
) (resp *depotpb.CreateShoppingListResponse, err error) {
    ctx = s.c.Scoped(ctx)
    defer func(tx *sql.Tx) {
```

```
        err = s.closeTx(tx, err)
    } (di.Get(ctx, "tx").(*sql.Tx))

    next := server{
        app: di.Get(ctx, "app").(application.App),
    }


    return next.CreateShoppingList(ctx, request)
}
```

Each of the other methods in serverTx work the same way:

1. Create a new scoped container in a new context.

2. Use a deferred function to commit or roll back the transaction from the scoped container.

3. Create a new server instance with a new scoped application from the container within the context.

4. Return as normal after calling the instanced server method with the context containing the scoped container from *step 1*.

The CreateShoppingList method makes use of named return values, used so that our transaction can be closed and committed or rolled back with this relatively simple method:

```
func (s serverTx) closeTx(tx *sql.Tx, err error) error {
    if p := recover(); p != nil {
        _ = tx.Rollback()
        panic(p)
    } else if err != nil {
        _ = tx.Rollback()
        return err
    } else {
        return tx.Commit()
    }
}
```

The transaction will be rolled back if there was a panic or an error was returned. We are not intending to catch panics here, so we will re-panic so that it can be recovered elsewhere up the stack. Otherwise, the transaction will be committed, and any error from that attempt will be returned as a new error. It is not happening here, but errors that result from rolling back the transaction could be logged.

## Updating the domain event handlers

The domain event handlers are a unique situation compared to the other handlers. They will be called on during requests that have been started by the gRPC server or other handlers. That means a scoped container will already exist in the context that the `handlers` function receives. Creating a new scoped container within the domain event handlers would mean we would also be creating and using all new instances of our dependencies. You can see an illustration of the process here:

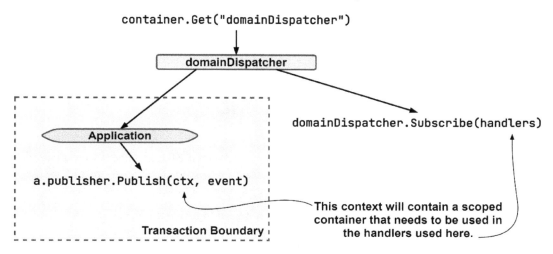

Figure 9.6 – Updating the domain dispatcher to use a scoped container

To make the preceding process work, `domainDispatcher` is registered as a singleton dependency. That way, the instance that is returned by any container will be the same instance regardless of scope. It also means we will be calling `Publish()` on the same instance that we had previously called `Subscribe()` on.

Then, in the `RegisterDomainEventHandlersTx()` function, we will need to use an anonymous function as our handler so that we can fetch an instance of `domainEventHandlers` for the current scope:

```
func RegisterDomainEventHandlersTx(
    container di.Container,
) {
    handlers := ddd.EventHandlerFunc[ddd.AggregateEvent](
        func(
            ctx context.Context,
            event ddd.AggregateEvent,
        ) error {
```

```
        domainHandlers := di.
            Get(ctx, "domainEventHandlers").
            (ddd.EventHandler[ddd.AggregateEvent])


        return domainHandlers.HandleEvent(ctx, event)
    })


    subscriber := container.
        Get("domainDispatcher").
        (*ddd.EventDispatcher[ddd.AggregateEvent])


    RegisterDomainEventHandlers(subscriber, handlers)
}
```

Inside the `handlers` anonymous function that we define, we do not use the container that was passed into `RegisterDomainEventHandlersTx()` or create a new scoped container. Instead, we use the `di.Get()` function to fetch a value from an already scoped container.

Later, when we implement the *Outbox pattern*, we will not need to revisit this function.

**Updating the integration event and command handlers**

Our updates to the integration event and command handlers will be like the gRPC `serverTx` updates. We want to define a transactional boundary and will need to start a new scope and transaction. Into a function named `RegisterIntegrationEventHandlersTx()`, we put the following updated event message handler:

```
evtMsgHandler := am.MessageHandlerFunc
    [am.IncomingEventMessage](
    func(
        ctx context.Context, msg am.IncomingEventMessage,
    ) error {
        ctx = container.Scoped(ctx)
        defer func(tx *sql.Tx) {
            // rollback or commit like in serverTx...
        }(di.Get(ctx, "tx").(*sql.Tx))
        evtHandlers := di.
            Get(ctx, "integrationEventHandlers").
            (ddd.EventHandler[ddd.Event])
```

```
            return evtHandlers.HandleEvent(ctx, msg)
    },
)
```

The command handlers work exactly like the integration event handlers, and the same updates can be applied there as well. A new anonymous function should be created that creates a new scope and fetches a scoped instance of `commandHandlers`.

At this point, the composition root has been updated to register all of the dependencies, and the application and handlers into a DI container. Then, the gRPC server and each handler receive some updates so that we have each request running within its own transactional boundary and with its own database transaction. Use of `di.Container` added a good deal of new complexity to the composition root in regard to managing our dependencies, but functionally, the application remained the same.

### Runs like normal

After making those changes, if you run the application now, there will be no noticeable change. There are no new logging messages to look for, and the **Depot** module will handle requests just as it did before, except now, with each request, a lot of new instances will be created to handle the requests that are then discarded when the request is done. Every query and insert will be made within a single transaction, effectively turning multiple writes into one. It will not matter which tables we interact with; the same transaction surrounds all interactions for each and every request now.

While we have made significant changes to the application and kept the observable functionality the same, the dual write problem has not been solved. Next, the Inbox and Outbox tables will be covered and then implemented to address the dual writes that exist in our application.

## Using an Inbox and Outbox for messages

We have now updated the `Depot` module so that we can work with a single database connection and transaction. We want to now use that to make our database interactions and message publishing atomic.

When we make the publishing and handling of the messages atomic alongside the other changes that are saved into our database, we gain the following benefits:

- **Idempotent message processing**: We can be sure that the message that we are processing will only be processed a single time
- **No application state fragmentation**: When state changes occur in our application, they will be saved as a single unit or not at all

With the majority of the work behind us to set up the transactions, we can now implement the inboxes and outboxes for our messages.

## Implementing a messages inbox

Back in *Chapter 6*, in the *Idempotent message delivery* section, I presented a way in which we could ensure that no matter how many times a message was received, it would only be handled one time.

The method presented was to use a table where the incoming message identity is saved, and if there is a conflict inserting the identity, then the message is not processed and simply acknowledged. When there is no conflict, the message is processed, and the identity of the message will be committed into the database along with the rest of the changes.

### Inbox table schema

We begin by looking at the table schema in which incoming messages will be recorded:

```
CREATE TABLE depot.inbox (
    id          text NOT NULL,
    name        text NOT NULL,
    subject     text NOT NULL,
    data        bytea NOT NULL,
    received_at timestamptz NOT NULL,
    PRIMARY KEY (id)
);
```

This table will hold every incoming `RawMessage` instance that the **Depot** module receives. In addition to being used for deduplication, it could also be used to replay messages. More advanced schemas could include aggregate versions or the publication timestamp to be used for a better ordering of messages as they are processed, and could also include the aggregate ID or the metadata in a searchable format so that the messages can be partitioned to scale message processing in the future. As it stands, this schema will be sufficient for our needs.

### Inbox middleware

Saving incoming messages will be handled with a middleware that will attempt to insert the message into the table as part of the deduplication process, as illustrated here:

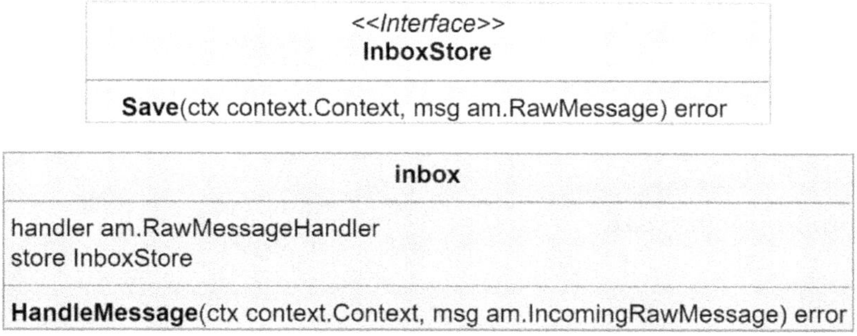

| <<Interface>> |
| --- |
| **InboxStore** |
| **Save**(ctx context.Context, msg am.RawMessage) error |

| inbox |
| --- |
| handler am.RawMessageHandler<br>store InboxStore |
| **HandleMessage**(ctx context.Context, msg am.IncomingRawMessage) error |

Figure 9.7 – The InboxStore interface and inbox middleware type

A factory for the inbox middleware is added to the container so that it can be injected into the handlers, as follows:

```
container.AddScoped("inboxMiddleware",
    func(c di.Container) (any, error) {
        tx := c.Get("tx").(*sql.Tx)
        inboxStore := pg.NewInboxStore("depot.inbox", tx)
        mw := tm.NewInboxHandlerMiddleware(inboxStore)
        return mw, nil
    },
)
```

We are now introducing a table into the message handling, and that means `inboxMiddleware` must become a scoped dependency. This dependency must be injected by every handler that subscribes to messages.

### Updating the handlers

The inbox middleware works with `IncomingRawMessages`, which our current streams, command, event, and reply do not handle. We will need to create new message handlers, which will work out because those streams are not scoped and because their subscribe sides cannot be scoped.

We can create a new `EventMessageHandler` instance, which does the work of the `EventStream` `Subscribe()` method but works with `RawMessages` instead:

```
type eventMsgHandler struct {
    reg     registry.Registry
    handler ddd.EventHandler[ddd.Event]
}
func NewEventMessageHandler(
```

```
        reg registry.Registry,
            handler      ddd.EventHandler[ddd.Event],
    ) RawMessageHandler {
        return eventMsgHandler{
            reg:      reg,
            handler: handler,
        }
    }
    func (h eventMsgHandler) HandleMessage(
        ctx context.Context, msg IncomingRawMessage,
        ) error {
        var eventData EventMessageData
        err := proto.Unmarshal(msg.Data(), &eventData)
        if err != nil { return err }
        eventName := msg.MessageName()
        payload, err := h.reg.Deserialize(
            eventName, eventData.GetPayload(),
        )
        if err != nil { return err }
        eventMsg := eventMessage{
            id:         msg.ID(),
            name:       eventName,
            payload:    payload,
            metadata:   eventData.GetMetadata().AsMap(),
            occurredAt: eventData.GetOccurredAt().AsTime(),
            msg:        msg,
        }
        return h.handler.HandleEvent(ctx, eventMsg)
    }
```

The new event message handler is brought together with other dependencies in an updated anonymous function inside of the `RegisterIntegrationEventHandlersTx()` function:

```
evtMsgHandler := am.RawMessageHandlerFunc(func(
    ctx context.Context,
    msg am.IncomingRawMessage,
    ) (err error) {
```

```
    ctx = container.Scoped(ctx)
    // existing rollback or commit code snipped...
    evtHandlers := am.RawMessageHandlerWithMiddleware(
        am.NewEventMessageHandler(
            di.Get(ctx, "registry").
                    (registry.Registry),
            di.Get(ctx, "integrationEventHandlers").
                    (ddd.EventHandler[ddd.Event]),
        ),
        di.Get(ctx, "inboxMiddleware").
                (am.RawMessageHandlerMiddleware),
    )
    return evtHandlers.HandleMessage(ctx, msg)
})
```

Let's note a few points about the new function:

1.  It is now a RawMessageHandlerFunc type and is no longer an EventMessageHandlerFunc type.

2.  A middleware function is used to apply inboxMiddleware, which works exactly like AggregateStoreWithMiddleware, which was used to add domain publishers and snapshot support.

3.  evtHandlers implements RawMessageHandler and not EventHandler[Event] now.

The subscriber that the subscriptions are made on will now be the *stream* and not an event stream or command stream, as shown here. This, again, is because our inbox middleware is not message-type aware:

```
subscriber := container.Get("stream").(am.RawMessageStream)
```

With this update done, the integration event messages will now be deduplicated, and copies of each processed message will be kept in the depot.inbox table. This process is automatically going to be part of our scoped request due to the work done in the previous section.

## Implementing a messages outbox

To implement the Transactional Outbox pattern, we will be splitting the existing publishing action into two parts. The first part will consist of saving the outgoing message into the database, and the second part will be implemented as a new processor that will receive or check for records that are written into the database so that it can publish them to where they need to go. The first part of the Transactional Outbox pattern is shown in *Figure 9.8*. The transaction that we are creating for each request will be used so that all changes from whatever work we have done, and messages, are saved atomically:

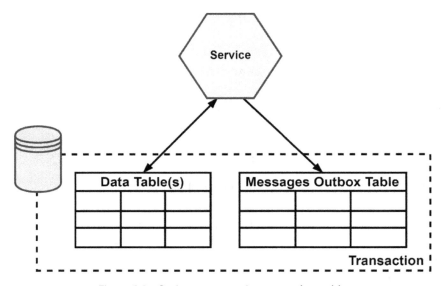

Figure 9.8 – Saving messages into an outbox table

### Outbox table schema

Starting the same way with the outbox as we did with the inbox, let's take a look at the table schema:

```
CREATE TABLE depot.outbox(
    id              text NOT NULL,
    name            text NOT NULL,
    subject         text NOT NULL,
    data            bytea NOT NULL,
    published_at timestamptz,
    PRIMARY KEY (id)
);
CREATE INDEX depot_unpublished_idx
```

```
ON depot.outbox (published_at)
WHERE published_at IS NULL;
```

This table is very similar to the `depot.inbox` table, with only a couple of differences:

- The `received_at` column is renamed `published_at`, and it also allows `null` values
- We add an index to the `published_at` column to make finding unpublished records easier

This table could also be updated to include more advanced columns such as the aggregate information, which could be used for ordering or partitioning.

## Outbox middleware

A middleware is created to catch outgoing messages to save them into the outbox table, as illustrated here:

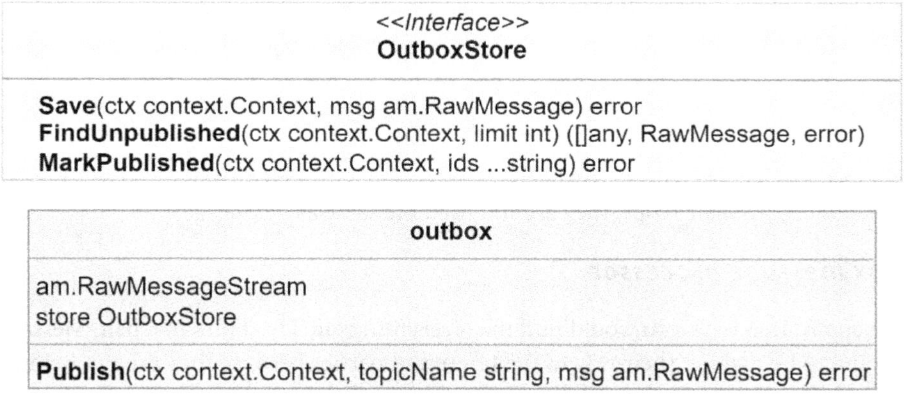

Figure 9.9 – The OutboxStore interface and outbox middleware type

The middleware this time will be for a `RawMessageStream` instance, and it will be used on a stream and not used on the handlers. A new scoped stream dependency is created to be used with the different types of streams that are being used:

```
container.AddScoped("txStream",
    func(c di.Container) (any, error) {
        tx := c.Get("tx").(*sql.Tx)
        outboxStore := pg.NewOutboxStore(
            "depot.outbox", tx,
        )
        return am.RawMessageStreamWithMiddleware(
            c.Get("stream").
```

```
                    (am.RawMessageStream),
            tm.NewOutboxStreamMiddleware(outboxStore),
        ), nil
    },
)
```

This new dependency will be used by each message-type stream in place of the original stream dependency. Here's an example for `eventStream`:

```
container.AddScoped("eventStream",
    func(c di.Container) (any, error) {
        return am.NewEventStream(
            c.Get("registry").(registry.Registry),
            c.Get("txStream").(am.RawMessageStream),
        ), nil
    },
)
```

The streams must now become scoped dependencies because they depend on other scoped dependencies. Outside of the change to their scope, they are still used the same as before.

### The outbox message processor

Running the application now, you would find that everything quickly comes to a halt. Messages are no longer going to be making their way to the interested parties because they are not being safely stored inside a local table for each module that has been updated to use the outbox table. We have only implemented one side of the pattern; to get the messages flowing once more, we will need to add a second side—a message processor.

Our implementation of the outbox message processor will use polling, but a more performant option would be to read the PostgreSQL **write-ahead log** (**WAL**), the write-ahead-log, as this method will not cause additional queries to be constantly run against the tables.

The following diagram illustrates the process:

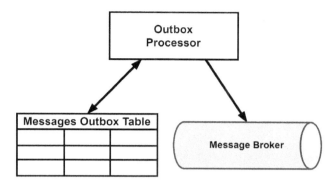

Figure 9.10 – Processing outbox messages

Our processor will fetch a block of messages, publish each of them, and then update the table to mark them as actually having been published. The processor itself suffers from a dual write problem, but when it fails, the result will be that one or more messages are published more than once. We already have deduplication in place thanks to our implementation of the inbox, so the modules will be protected from any processor failures.

As with the saga orchestrator, an Outbox message processor is a process that can live just about anywhere. It can have its own services that are designed to scale horizontally, making use of whatever partitioning logic is necessary. In our application, the processors will be run as another process within all of the existing modules, as illustrated here:

Figure 9.11 – The outbox processor interface and struct

Each processor is given singleton streams and database connections, as shown in the following code snippet. We do not want to use transactions because this will be running continuously:

```
container.AddSingleton("outboxProcessor",
    func(c di.Container) (any, error) {
        return tm.NewOutboxProcessor(
            c.Get("stream").(am.RawMessageStream),
            pg.NewOutboxStore(
```

```
                    "depot.outbox",
                    c.Get("db").(*sql.DB),
                ),
            ), nil
        },
    )
```

This new dependency is used in a goroutine so that it runs alongside all the other servers and handlers:

```go
func startOutboxProcessor(
    ctx context.Context, container di.Container,
) {
    outboxProcessor := container.
        Get("outboxProcessor").
            (tm.OutboxProcessor)
    logger := container.Get("logger").(zerolog.Logger)
    go func() {
        err := outboxProcessor.Start(ctx)
        if err != nil {
            logger.Error().
                Err(err).
                Msg("depot outbox processor encountered
                    an error")
        }
    }()
}
```

For our application and for demonstration purposes, the processor will fetch up to 50 messages at a time to publish and will wait for half a second in between queries looking for messages that need to be published; it does not wait at all to fetch new messages if it just published some. A more robust outbox processor would allow the number of messages and the time to wait before looking for messages to be configurable.

# Summary

In this chapter, we made more changes to the composition roots than ever. We used a small DI package to store value factories and to fetch new instances as needed. We are also able to fetch instances that are scoped to each message or request our application receives. We also implemented a messages deduplication strategy using an inbox table.

The Transactional Outbox pattern was also implemented along with local processes to publish the messages stored in an outbox table. As a result of these updates, the reliability of messages arriving at their destinations when they should and the risk of making incorrect updates as a result of reprocessing a message has been reduced a considerable amount. The event-driven version of MallBots has become a very reliable application that is much more resilient to problems springing up compared to the original synchronous version of the application.

In the next chapter, we will cover testing. We will develop a testing strategy that includes unit testing, integration testing, and E2E testing. I will also introduce you to contract testing, a relatively unknown form of testing that combines the speed of unit tests with the test confidence of large-scope integration testing. We will also discuss additional test topics such as table-driven testing, using test suites, and more.

# Part 3:
# Production Ready

In this last part, we will cover the topics of testing, deployment, and observability. We will begin by discussing testing strategies and going over the different kinds of tests we can use to ensure our application works as intended. Next, we will refactor the application from a modular monolith into microservices that can be deployed into a cloud environment. Then, we will update the application so that it can be monitored using logging, metrics, and distributed traces.

This part consists of the following chapters:

# 10
# Testing

In *Part 2* of this book, we took an entirely synchronous application and transformed it into an asynchronous application using events and messaging. Our application is more resilient and agile but has gained some new libraries and dependencies as a result.

Testing an asynchronous application can pose some unique challenges but remains within reach by following testing best practices. In this chapter, we will look at testing the MallBots application from the unit test level and writing executable specifications using the Gherkin language.

In this chapter, we will cover the following topics:

- Coming up with a testing strategy
- Testing the application and domain with unit tests
- Testing dependencies with integration testing
- Testing component interactions with contract tests
- Testing the application with end-to-end tests

## Technical requirements

You will need to install or have installed the following software to run the application or to try this chapter's examples:

- The Go programming language version 1.18+
- Docker

The code for this chapter can be found at https://github.com/PacktPublishing/Event-Driven-Architecture-in-Golang/tree/main/Chapter10.

# Coming up with a testing strategy

For applications such as MallBots, we should develop a testing strategy that tests whether the application code does what it is supposed to be doing. It should also check whether various components communicate and interact with each other correctly and that the application works as expected:

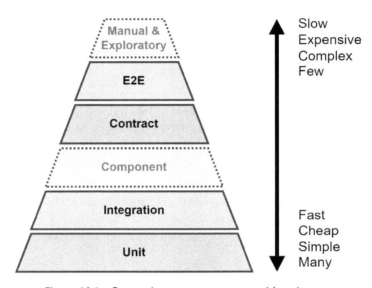

Figure 10.1 – Our testing strategy as a pyramid or ziggurat

Our testing strategy will have four parts:

- Unit tests
- Integration tests
- Contract tests
- End-to-end tests

Unit tests are a no-brainer addition to our strategy; we want to ensure the code we write does what we intend it to do. We want to test the input and output from the module core and include an integration test to test the dependencies that it uses. We will use contract tests to detect any breaking changes to the application's many APIs and messages that tie the modules together. Finally, we want to run tests that check that the application is functioning as per stakeholder expectations and will use **end-to-end** (**E2E**) testing.

There are additional levels and forms of testing that we could include, such as component testing. This would be used to test each module in isolation – that is, like an E2E test but limited to just that module. We may also see some manual tests take place or have the testing or development teams work through scenarios to perform exploratory testing. We could also stress or load test the application, which could be added to the strategy later as the application matures.

## Unit tests

Unit tests should make an appearance in any testing strategy. They are used to test code for correctness and to locate problems with application and business logic implementations. In a testing strategy, they should take up the bulk of the testing efforts. These tests should be free of any dependencies, especially any I/O, and make use of test doubles such as mocks, stubs, and fakes. The system under test for a unit test should be very small; for example, individual functions and methods.

> **System under test**
>
> At each level of testing, we use the term **system under test** (**SUT**) to describe the component or components being tested. For unit tests, the SUT may be a function, whereas for E2E testing, it would encompass the application and any external APIs involved. Generally, the SUT expands in scope or application coverage the higher up you go in the testing pyramid.

Any application can benefit from having unit tests in its testing strategy. If used sparingly, extremely fast-running tests can focus on logic and algorithms that are complex or critical to the success of the business.

## Integration tests

Next up is integration testing where, instead of focusing on the logic, you will focus on testing the interactions between two components. Typically, you must test the interactions between a component with one of its dependencies. Testing that your ORM or repository implementations work with a real database would be an example of an integration test. Another example would be testing that your web interface works with application or business logic components. For an integration test, the SUT will be the two components with any additional dependencies replaced with mocks.

Applications with complex interactions in their infrastructure can benefit from the inclusion of integration tests in the testing strategy. Testing against real infrastructure can be difficult or too time-consuming, so teams may decide to not do so or only develop a few critical path tests.

## Contract tests

A distributed application or a modular monolith like ours is going to have many connection points between the microservices or modules. We can use contract tests built by consumers' expectations of an API or message to verify whether a producer has implemented those expectations. Despite being rather high on the testing pyramid, these contract tests are expected to run just as fast as unit tests since they do not deal with any real dependencies or test any logic. The SUT for a contract will be either the consumer and its expectation, or the producer and its API or message verification.

Distributed applications will benefit the most from adding contract tests to the testing strategy. These tests are not just for testing between microservices – they can also be used to test your UI with its backend API.

## End-to-end tests

E2E tests are used to test the expected functionality of the whole application. These tests will include the entire application as the SUT. E2E tests are often extensive and slow. If your application includes a UI, then that too will become part of the tests because they intend to test the behaviors and correctness of the application from the point of view of the end user. The correctness being tested for is how the application performs and not like a unit test's correctness of how the application does it.

Teams that take on the effort of maintaining fragile and costly tests are rewarded with confidence that the application is working as expected during large operations that can span the whole application.

In the upcoming sections, we will explore each of the testing methods present in our testing strategy.

# Testing the application and domain with unit tests

The system under test for a unit test is the smallest unit we can find in our application. In applications that are written in Go, this unit will be a function or method on a struct:

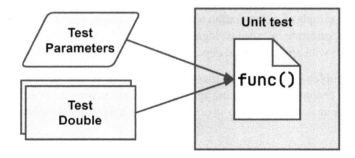

Figure 10.2 – The scope of a unit test

As shown in *Figure 10.2*, only the function code is being tested. Any dependencies that the code under test requires must be provided as a test double such as a mock, a stub, or a fake dependency. Test doubles will be explained a little later in the *Creating and using test doubles in our tests* section.

Each test should focus on testing only one path through the function. Even for moderately complex functions, this can result in a lot of duplication in your testing functions. To help with this duplication, the Go community has adopted table-driven tests to organize multiple tests of a single piece of code under test into a single test function.

## Table-driven testing

This method of testing was introduced to the Go community by Dave Cheney in his similarly named blog post, *Prefer table driven tests* (https://dave.cheney.net/2019/05/07/prefer-table-driven-tests). Table-driven tests are made up of two parts – a table of test cases and the testing of those test cases.

### The table of test cases

A slice of structs that contains the test inputs and outputs is called a test case. The following listing shows a table with two test cases using a map to build the table:

```
tests := map[string]struct {
    input string
    want  int
}{
    "word":  {input: "ABC", want: 3},
    "words": {input: "ABC ABC", want: 6},
}
```

If we used a slice instead of the map, then we would want to include an additional field in the struct to hold a string that is used as the subtest's name.

### Testing each test case

The actual testing will depend on how the unit needs to be tested. However, there is some simple boilerplate code that we should use so that we can make sense of the test failures, should they pop up.

In the following code block, the highlighted code is the simple boilerplate that is used to run through each test case:

```
for name, tc := range tests {
    t.Run(name, func(t *testing.T) {
        // arrange, act, and assert
```

```
    })
}
```

In the loop, we use the **subtesting** feature to run each test case under the heading of the original test function name. The following output is an example of running the `AddItem` application tests for the **Shopping Baskets** module:

```
--- PASS: TestApplication_AddItem (0.00s)
    --- PASS: TestApplication_AddItem/NoBasket (0.00s)
    --- PASS: TestApplication_AddItem/NoProduct (0.00s)
    --- PASS: TestApplication_AddItem/NoStore (0.00s)
    --- PASS: TestApplication_AddItem/SaveFailed (0.00s)
    --- PASS: TestApplication_AddItem/Success (0.00s)
PASS
```

The `AddItem` test has five test cases that test how the input to the function might be handled under different conditions. This test can be found in the `/baskets/internal/application/application_test.go` file.

The application that `AddItem` is defined on has several dependencies, and each of those is replaced with test doubles so that we can avoid dealing with any real I/O. We also want to intercept calls into the dependencies to control which path through the `AddItem` method we are testing.

We will want to use a test double that is not only able to intercept the calls but also able to send back programmed responses. There are several kinds of test doubles, so let's look at them and see which works best for us here.

## Creating and using test doubles in our tests

Test doubles are tools we can use to isolate the system or code under test from the rest of the system around it.

These tools come in different forms, each useful for different testing scenarios:

- Fakes implement the same functionality as the real dependency. An in-memory implementation of a repository could stand in and take the place of a PostgreSQL implementation so that the test does not rely on any real I/O.

- Stubs are like fakes, but the stub implementation responds with static or predictable responses.

- Spies work like an observable proxy of the real implementation. A spy can be used to report back the input, return values, and the number of calls that it received. Spies may also help with recording the inputs and outputs that were seen for later use.

- Mocks mimic the real implementation, similar to a fake, but do not provide an alternative implementation. Instead, a mock is configured to respond in certain ways to specific inputs. Then, like a spy, it can be used to assert whether the proper inputs were received, the right number of calls were made, and no unexpected calls or inputs were encountered.

Fakes and stubs can be used when the interaction with the dependency is not important to the test, whereas spies and mocks should be used when the input and responses matter.

### Working with mocks

For our unit test, we will use mocks. To create the mocks that we will use, we will use the Testify mocks package (`https://github.com/stretchr/testify`). This will provide the mocking functionality, along with the mockery tool (`https://github.com/vektra/mockery`) to make generating them a breeze. The mockery tool can be installed with the following command:

```
go install github.com/vektra/mockery/v2@v2.14.0
```

Each module that will be tested using mocks will have the following line added to its `generate.go` file; for example, `/baskets/generate.go`:

```
//go:generate mockery --all --inpackage --case underscore
```

This `go:generate` directive will look for the interfaces defined within the directory and subdirectories and create mocks of them. `--inpackage` and `--case underscore` will configure the tool to create the mocks in the current package using underscores in the filename. The `--all` flag will make the tool generate a mock for each interface that is found. When mockery creates mocks next to the interface, it will add a `Mock` prefix to each interface that it found in a file with a prefix of `mock_`. For example, the `Application` interface is mocked as `MockApplication`, and that mock will be found in `/baskets/internal/application/mock_application.go`.

Organizing and naming test doubles comes down to preferences in most cases. My preference is to place them next to the interfaces and implementations that they double. Another preference is to keep the naming simple and use either a prefix or suffix to identify the type of test double that you are dealing with.

With the mocks created, we need to use them in our tests. To do that, we will include a new field in our test case structs so that they can be configured for each test case:

```
type mocks struct {
    baskets   *domain.MockBasketRepository
    stores    *domain.MockStoreRepository
    products  *domain.MockProductRepository
    publisher *ddd.MockEventPublisher[ddd.Event]
}
```

```
tests := map[string]struct {
    ...
    on         func(f mocks)
    wantErr bool
}{...}
```

In the previous listing, a named `struct` has been created with field types of the actual mocks; using the interfaces here will not help us since we want the concrete mock implementations. Then, in the anonymous struct that defines our test cases, we have added a function that accepts the mocks struct. With these additions, any test case that is expected to make calls into a mock can do so during the **Arrange** portion of the test function.

> **Arrange, act, and assert**
>
> The **Arrange-Act-Assert** (**AAA**) pattern is a simple yet powerful way to build your tests. It breaks up a test function into three parts for better maintainability and readability. The Arrange portion is where the test is set up, the Act portion is where the target is called or interacted with, and the Assert portion is where the final test or verification happens. By following this pattern, it is easy to spot test functions that are doing more than testing one thing at a time. This pattern is also known as Given-When-Then.

In the test function, we must set up the mocks, execute the method, and perform our assertions using the following code:

```
// Arrange
m := mocks{
    baskets:   domain.NewMockBasketRepository(t),
    stores:    domain.NewMockStoreRepository(t),
    products:  domain.NewMockProductRepository(t),
    publisher: ddd.NewMockEventPublisher[ddd.Event](t),
}
if tc.on != nil {
    tc.on(m)
}
a := New(m.baskets, m.stores, m.products, m.publisher)

// Act
err := a.AddItem(tc.args.ctx, tc.args.add)

// Assert
```

```
if (err != nil) != tc.wantErr {
    t.Errorf("AddItem() unexpected error = %v", err)
}
```

During each test case run, we will create new mocks but leave them alone if no function has been defined to configure them. A mock will fail the test if it is called and we have not configured any calls. This will be helpful because we do not need to remember which calls or mocks have been set up, nor which to change when we make changes to the code under test.

The Mockery package has generated constructors for our mocks that accept the test variable. Using the constructors, we do not need to include any additional assertions for the Assert portion of the test function. When the test completes, each mock will be automatically checked to ensure that the exact number of calls were made into it and that the calls included the correct inputs:

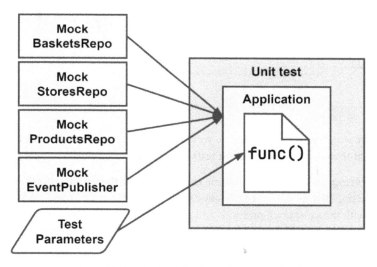

Figure 10.3 – The SUT for the AddItem method

To test the AddItem method on Application, we must provide an application instance with all the dependencies that it needs and then pass additional parameters to the AddItem method. The method only returns an error, so using a mock double instead of any of the others makes the most sense. Without mocks, we would not be able to see into the method.

## Testing dependencies with integration testing

An application is made up of many different components; some of those components are external to the application. These external components can be other services, or they can be infrastructure that needs to be in place for the application to function properly.

It is hard to find any application built for the web that does not interact with infrastructure. Actually, it's impossible – the web server or API gateway that the application would use to receive requests would fall into the definition of infrastructure.

We cannot test these dependencies using a unit test because if we replaced the dependency with any kind of test double, then it would not be a true test of that dependency.

In an integration test, both components are the SUT, which means we need to test with real infrastructure when possible:

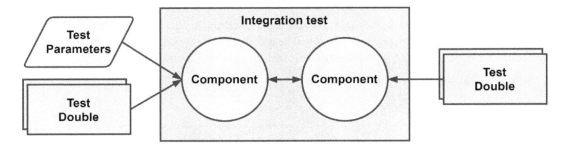

Figure 10.4 – The scope of an integration test

Unlike the unit tests, which were expected to be very simple, at least in terms of what the environment must provide so that they can run, integration tests ramp up the complexity a great deal.

A wrong way to develop these tests is to assume each developer that might run them has the same environment as you do. That would mean the same OS, the same installed tools, and the same locally accessible services with the same configurations and permissions. You would then write the test leaving all of those environmental expectations out and leaving out how to run the tests in the documentation or some other form of knowledge share.

A better way to write integration tests would be to bring what is necessary to run the test into that test without requiring any test-specific environment and setup.

## Incorporating the dependencies into your tests

A lot of the services and infrastructure in use today are available as a **Docker** container; this could be a real production-ready container such as the ones for many databases. Some containers are purpose-built to aid in development or testing efforts. An example of a container that can help with development would be **LocalStack** (https://localstack.cloud), a container that provides a local development and testing environment for many of the offerings from AWS.

Using Docker and containers is a great way to bring these dependencies into your local environment. However, the challenge in using them in tests is that we want to be able to control their state and may also want to set them up in different ways to support different tests. However, we need to know how to incorporate these containers into our tests.

### Option 1 – manually by using Docker Compose files

We can create a Docker Compose file for our tests, such as `test.docker-compose.yml`, that will stand up everything we will need to connect to for the integration tests that we'll write. This should make it easy for every developer to have the dependencies available, and so long as everyone remembers to start up the environment, they should also have no issues running the tests. Volumes can be destroyed during the environment teardown so that previous runs do not affect others.

The downsides of this option begin with the Compose file itself. If a problem exists when standing up the entire environment, then someone will need to make changes to it before they can test. There may also be issues running the tests multiple times, so tearing down the environment to stand it back up again might be necessary, which could take a considerable amount of time. To tackle this, we can take a different approach.

### Option 2 – internalizing the Docker integration

There is a solution we can use that will not only allow us to use different containers or compose environments for different tests but also remove the step of having to run a Docker command before executing any integration tests.

`Testcontainers-Go` (`https://golang.testcontainers.org`) is a library that makes it possible to start up a container or compose an environment that is controlled by code that we can include in our tests.

The benefits of this option are that we will always have a pristine environment to run our tests in and subsequent runs will not need to wait for any containers or volumes to be reset. The other is the containers will always be started and removed when the test is run. This means that there is no longer any need to maintain documentation on how to prepare a local environment to run tests. This is the better option in my opinion, but it will require some initial setup, as well as some resetting or cleanup between each test.

## Running tests with more complex setups

Our integration tests will likely end up being a little more complex than the unit tests we have previously worked with. We may require certain actions to occur at the start of the run and the end; likewise, we need actions to run before and after each test. This is not a difficult task by any stretch. We can write the test harness ourselves, but whatever we write should also contain tests. Instead, we can use an existing test harness that handles all of this for us.

This harness is the Testify suite package. When we are using this new harness, we can continue to use table-driven tests, but we need to manage the state setup and reset ourselves.

To start using Testify suites, create a new struct and include `suite.Suite` as an anonymous field. Then, create a simple test function to run the suite:

```
type productCacheSuite struct {
    suite.Suite
    // [optional] any additional fields
}

func TestProductCacheRepository(t *testing.T) {
    suite.Run(t, &productCacheSuite{})
}
```

We can include additional fields in the struct that can be accessed by the test methods.

## Testing ProductCacheRepository

We will use all of the aforementioned methods to test the interaction between the PostgreSQL implementation of `ProductCacheRepository` and PostgreSQL:

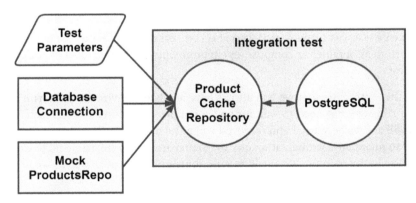

Figure 10.5 – The integration test for ProductCacheRepository

This implementation uses a connection to the database and also has a dependency on the `ProductRepository` interface. In the application, this is implemented as a gRPC client, which will fetch a `Product` instance when we cannot find it in the database. For this integration test, that dependency will be mocked. Before we write our first test, we need to configure the suite so that our tests can use a real database and still be isolated from each other.

## Suite composition

Inside the `productCacheSuite` struct, we will add the following additional fields:

- `container`, which will hold the reference to the PostgreSQL container we have started.

- `db`, which will be a real database connection to PostgreSQL. We will use it to reset the database in between tests.

- `mock` will be an instance of `MockProductRepository`. If we had other dependencies to mock or fake, we would have used a less generic name.

- `repo`, which is a real instance of the PostgreSQL implementation that we intend to test.

These fields will be accessible to our tests, as well as to the methods we will use to set up the suite and each test.

## Suite setup

We must start by setting up the suite with some fields we want to make available to all tests. The first of those is the database connection. To make that connection, we need to have a database we can connect to. The following listing is how the PostgreSQL container is started up:

```
const dbUrl = "postgres://***:***@localhost:%s/mallbots"
s.container, err = testcontainers.GenericContainer(ctx,
    testcontainers.GenericContainerRequest{
        ContainerRequest: testcontainers.ContainerRequest{
            Image:        "postgres:12-alpine",
            ExposedPorts: []string{"5432/tcp"},
            Env: map[string]string{
                "POSTGRES_PASSWORD": "***",
            },
            Mounts: []testcontainers.ContainerMount{
                testcontainers.BindMount(
                    initDir,
                    "/docker-entrypoint-initdb.d",
                ),
            },
            WaitingFor: wait.ForSQL(
                "5432/tcp",
                "pgx",
                func(port nat.Port) string {
```

```
                          return fmt.Sprintf(
                              dbUrl,
                              port.Port(),
                          )
                      },
                  ).Timeout(5 * time.Second),
              },
              Started: true,
          },
      )
```

This listing will start up a new container from the `postgres:12-alpine` image. Like the service entry in the `docker-compose.yml` file, we must give it a hostname and initialize it with some files we will mount in the container.

The `WaitingFor` configuration is used to block the startup process until the database is truly ready for requests. In the Compose file, we use a similar effect with a small `wait-for` script.

The `testcontainers-go` package can also stand up services defined in a Docker Compose file. We will not be making use of that feature, but you can learn more about it at `https://golang.testcontainers.org/features/docker_compose/`.

Once the container is running and we are waiting for it to become available, we can make that database connection.

### Test setup

Before each test, a new mock is created, which is then injected along with the database connection into `ProductCacheRepository`:

```
func (s *productCacheSuite) SetupTest() {
    s.mock = domain.NewMockProductRepository(s.T())
    s.repo = NewProductCacheRepository(
        "baskets.products_cache",
        s.db,
        s.mock,
    )
}
```

We are keeping a reference to the mock because we will want to configure it during tests to expect calls. If we did not keep a reference, there would be no way to configure it from `ProductCacheRepository`.

## Test teardown

Every test should have the same slate upon which it will run. In our tests for the database, we will be creating new rows, updating rows, or deleting them. Without resetting the database in between each test, we may find ourselves in situations where the order of the tests affects subsequent tests passing or failing:

```
func (s *productCacheSuite) TearDownTest() {
    _, err := s.db.ExecContext(
        context.Background(),
        "TRUNCATE baskets.products_cache",
    )
    if err != nil {
        s.T().Fatal(err)
    }
}
```

We will keep things simple and TRUNCATE any tables that we work with. This is safe if this test suite is always using a PostgreSQL container that exists only for this test suite.

## Suite teardown

When all of the tests have finished running, we no longer need the connection to the database. The container should also be cleaned up and removed:

```
func (s *productCacheSuite) TearDownSuite() {
    err := s.db.Close()
    if err != nil {
        s.T().Fatal(err)
    }
    err := s.container.Terminate(context.Background())
    if err != nil {
        s.T().Fatal(err)
    }
}
```

In reverse order from what happened in the SetupSuite() method, we close the database connection and then terminate the container, which removes it and any volumes we might have created.

## The tests

With all the setup and teardown taken care of, our tests are going to be simple and to the point, much like the unit tests were. The following listing shows the test for the rebranding functionality:

```go
func (s *productCacheSuite) TestPCR_Rebrand() {
    // Arrange
    _, err := s.db.Exec("INSERT ...")
    s.NoError(err)
    // Act
    s.NoError(s.repo.Rebrand(
        context.Background(),
        "product-id",
        "new-product-name",
    ))
    // Assert
    row := s.db.QueryRow("SELECT ...", "product-id")
    if s.NoError(row.Err()) {
        var name string
        s.NoError(row.Scan(&name))
        s.Equal("new-product-name", name)
    }
}
```

We can access any of the fields defined in the suite and can even organize the tests in AAA fashion. During the Arrange phase of this test, we use the database connection to insert a new product cache record that is then acted upon in the next phase. The suite also has access to all the usual Testify assert functions, and we can skip importing that package in favor of using the assertion methods directly from the suite itself.

## Breaking tests into groups

Integration tests do not need to run quickly, and for good reason. Integration tests will typically need to deal with I/O, which is not exactly fast or predictable. Skipping or excluding the longer-running tests will be necessary if you want to keep the wait for test feedback as low as possible when developing some new logic or feature.

There are three ways to break long-running tests into groups or exclude them when running fast-running unit tests.

## Running specific directories, files, or tests

You can specify specific files, directories, and even individual tests when using the `test` command. This option will not permanently break your tests up into different groups that can be run separately, but outside of using your IDE, it presents the easiest way to target individual tests.

To run all the application tests for the **Shopping Baskets** module, you would use the following command:

```
go test ./baskets/internal/application
```

To run only the `RemoveItem` test, you would add `-run "RemoveItem$"` to the command:

```
go test ./baskets/internal/application -run "RemoveItem$"
```

We can target specific table-driven subtests as well. To run only the `NoProduct` subtest for the `RemoveItem` test, we can use `"RemoveItem/NoProduct$"`. For the following command, I have moved into the internal directory:

```
go test ./application -run "RemoveItem/NoProduct$"
```

In the previous two command examples, I used a Regex to search for the test to run. You can target a group of tests with a well-written Regex. The `test` tool makes it very easy to target specific tests when we need to be very focused on a test or a collection of tests.

## Go build constraints

We can use the conditional compilation build constraints to create groups of our tests. These constraints are normally used to build our programs for different OSs or CPU architectures, but we can also use them for our tests because the tests and the application are both compiled when we run the tests. Because this is accomplished by adding a special comment to the top of our files, we can only group tests together by files; we cannot create any subgroups of the tests within the files.

To group tests into an `integration` grouping, we can add the following with a second blank line to the top of the tests file:

```
//go:build integration
```

The following are a few rules that need to be followed for the compiler to recognize the comment as a build constraint:

- There must not be any spaces between the single-line comment syntax and `go:build`. Multiline comment syntax will not work.
- The constraint must be followed by a blank line.
- The constraint must be the first line in the file.

The file will now be ignored when we run the test command. The examples from the previous section would all ignore the tests in this file, even if we were to target the file and tests specifically. To run the tests now, we will need to pass the -tags option into the test command, like so:

```
go test ./internal/postgres -tags integration
```

You can combine multiple tags to create subgroups using the build constraints by taking advantage of the *Boolean* operators that it supports. We can modify the constraint so that the database tests are run with all integration tests or can be run by themselves:

```
//go:build integration || database
```

A file with this constraint could be run using any of the following commands:

```
go test ./internal/postgres -tags integration
go test ./internal/postgres -tags database
go test ./internal/postgres -tags integration,database
```

Using build constraints is a powerful and easy way to create groups of tests. Without the -tags option, any file that uses a build constraint will be ignored. This also has the downside of potentially skipping tests that are broken and not knowing it. The constraints at the top of the file can include typos or logical errors caused by incorrect operator usage.

When using build constraints, it is best to keep it simple.

### Using the short test option

The final method we will look at is the short test mode, which is built into the test tool. To enable short mode, you can simply pass in the -short option to any test command that you run. By itself, nothing happens, but if you include a check in your tests, you can exclude the longer-running tests from running. The test tool itself is not able to determine which tests are long-running tests; that determination is up to you.

We can skip long-running tests by using a block of code such as this:

```
func TestProductCacheRepository(t *testing.T) {
    if testing.Short() {
        t.Skip("short mode: skipping")
    }
    suite.Run(t, &productCacheSuite{})
}
```

The entire suite of tests will be skipped when the following command is run:

```
go test ./internal/postgres -tags integration -short
```

Checking for short mode can be added to any individual tests and subtests as well; we do not need to limit ourselves to tests run via a suite. Skipping tests with short mode allows us to be more selective about which test or tests are ultimately skipped.

The downside to using short mode is that the long-running tests are included by default, and we need to enable short mode to skip them. Another downside is that the option can be either on or off; there is no way to split your tests into more than two groups.

All three options I've mentioned can be used together. You could treat short mode as a way to skip tests that are just a little longer when running unit tests, and likewise for the other kinds of tests when used with the -tags option.

By using Docker containers, we can test more of our application by including real infrastructure in our tests, and by grouping the tests, we can exclude them when we want to run very fast unit tests. This form of testing will be too fragile to test integrations much larger than infrastructure interactions. For that testing, we can turn to contract tests.

## Testing component interactions with contract tests

We have chosen to create our application using the modular monolith pattern. Here, we have each module communicate with the others using either gRPC or by publishing messages into a message broker. These would be very common to see on any distributed application but would be rare or not used at all on a monolith or small application. What's more common to see across all applications is the REST API we use. This demonstration application does not have any true UI, but we have the API to support one. This API represents a third form of communication in our application, which is between an API provider and the API consumer.

We could test these interactions using integration tests since the definition of what an integration test covers is testing the interactions or integration between two components. However, the integration tests we wrote before tested smaller components, and the scope for the system under test was not very large. They are larger than the unit tests before them but are still small:

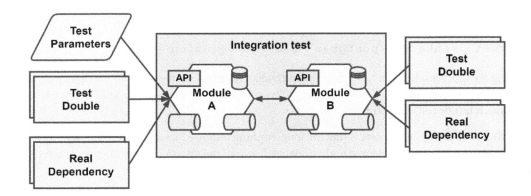

Figure 10.6 – System under test for an integration test of two modules

Testing even two modules together using an integration test would be a very large jump regarding how much of the system under test would now be forced into the scope of the test. There may be several real dependencies that are too difficult to replace with a test double and a real dependency would need to be stood up and used for the test.

Another possible but very likely issue with using integration tests in this manner is that we could be testing two components that have entirely different development teams and release schedules.

We want to minimize the extraneous components that get included in the test scope. This means we should only target the REST API if that is what we are interested in testing. The same goes for messaging; we should test whether we are receiving the messages that we expect and leave the rest of the module out of the equation. Contract testing allows us to focus on the APIs between providers and consumers, just like the integration tests do, but it allows us to run the tests in isolation, similar to a unit test.

Contract testing comes in two forms:

- **Consumer-driven contract testing (CDCT)**, which is when the contract is developed using the expectations of the consumers

- **Provider-driven contract testing (PDCT)**, which is when the contract is developed using the provided API of the provider

We will be using CDCT for our testing:

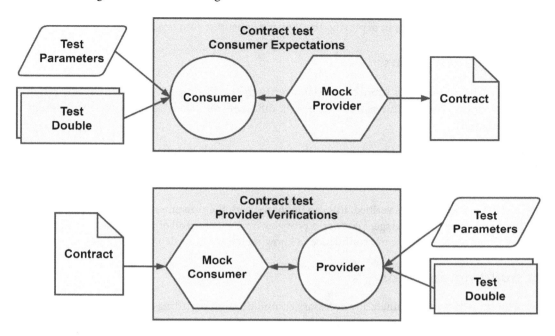

Figure 10.7 – System under test for consumer-driven contract testing

Contract testing is broken down into two parts: consumer expectations and provider verifications. Between the two sits the contract that is produced by the consumer when using consumer-driven testing.

On the consumer side, real requests are made to the mock provider, which will respond with simulated responses. On the provider side, the real requests will now be used as simulated requests and the provider will respond with real responses that are verified against the expected responses recorded in the contract.

Because the consumer is creating expectations, there would be no value in only running the consumer side without the provider verifying those expectations. Each side, both consumer and provider, has different contract testing goals.

## Consumer expectations

The consumers of an API will uniquely use that API. This could mean that it uses a fraction of the provided API endpoints or messages, and it could also mean that it is using only a portion of the data that it is provided.

Consumers should write their expectations based on what they use. This allows providers that are tested with contract testing to know what endpoints and data are being used by the consumers.

Consumers' expectations will change over time, as will the contracts. Processes can be set up in your CI/CD pipeline so that these changed contracts can be automatically verified with the provider to ensure that there are no issues in deploying the updated consumer into production.

## Provider verifications

Providers will be given one or more contracts to verify their API support. Each contract that they receive will expect different things from their API, different collections of endpoints, or different simulated requests.

The providers will be expected to implement the tests to verify the simulated requests against their real API. However, they may use whatever test doubles they need so that they don't have to stand up their entire module or microservice.

When a consumer's contract is verified, this can be shared with the consumer so that they know it will be OK to deploy with its API usage. Likewise, a provider, having passed all of the contract verifications it was presented with, will have the confidence in knowing it too can be deployed without any issues.

## Not building any silos

Contract testing does not eliminate any necessary communication regarding the integrations between teams; it helps them know about and get to the issues quickly. With contract testing, we achieve a high level of confidence on both sides that the integration is working. When issues are discovered during verification, then it is expected that the teams will have some dialog. Consumers can make mistakes and have incorrect expectations, which could mean there is room to improve or add API documentation. Providers may make a breaking change and will need to cooperate with the affected consumers to coordinate updates and releases.

## Contract testing with Pact

Just like using the Testify suite package for our more complex test setups, we will use a tool called Pact (`https://pact.io`) to handle a lot of the concerns outside of our tests. Pact provides libraries for many languages, which is handy for testing a JavaScript UI with your Go backend. Several tools can be used locally by the developers, as well as in the CI/CD process, to provide the promised confidence that deployments can happen with any issues.

### *Pact Broker*

Pact Broker (`https://docs.pact.io/pact_broker`) is an application we can start up in our environment to share contracts, as well as provide feedback for consumers stating that their contracts have been verified by the provider:

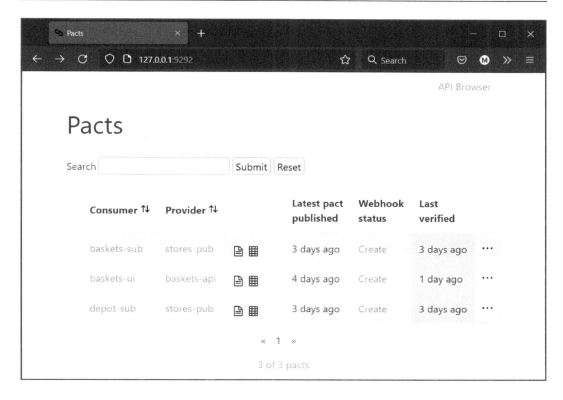

Figure 10.8 – Pact Broker showing our example integrations

Pact Broker can also be integrated with your CI/CD process to automate the testing of providers when a consumer has created or updated a contract. Likewise, consumers can be automatically tested when a provider has made changes to their API:

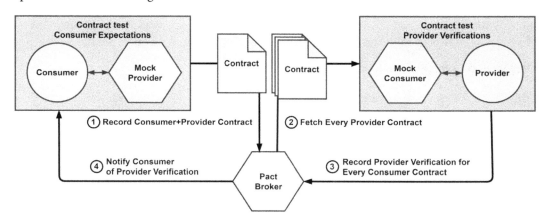

Figure 10.9 – Contract creation and verification flow using Pact Broker

Pact Broker may be installed locally using a Docker image, though you may use the hosted version with a free account at `https://pactflow.io/`.

### CLI tools

Pact will take care of creating and running the mock provider and consumer, but this functionality will require the necessary Pact CLI tools to be installed and available (`https://docs.pact.io/implementation_guides/cli`). You may choose either a Docker image or a Ruby-based standalone version.

### Additional Go tools

The provider example for the asynchronous tests uses an updated version of the Go libraries. If you would like to follow along and run these tests, you will need to install the `pact-go` installer and use it to download some additional dependencies:

```
go install github.com/pact-foundation/pact-go/v2@2.x.x
pact-go -l DEBUG install
```

The two preceding commands will download some files that will allow the updated provider verifications to run.

> **pact-go versioning**
>
> At the time of writing this book, the version used was tagged as `v2.x.x`. The minor and patch version values are `x`.

## REST consumer and provider example

First, we will test a simple JavaScript client against the REST API provided by the **Shopping Baskets** module. We do not have a real UI to add tests to, but we can create a small JavaScript client library. For contract testing, we would only want to work with the client library anyhow, so this is not a big problem.

We will focus on a couple of endpoints for this demonstration:

```
const axios = require("axios");

class Client {
  constructor(host = 'http://localhost:8080') {
    this.host = host;
  }

  startBasket(customerId) {
```

```
    return axios.post(
        `${this.host}/api/baskets`,
        {customerId}
    )
  }


  addItem(basketId, productId, quantity = 1) {
    return axios.put(
        `${this.host}/api/baskets/${basketId}/addItem`,
        {productId, quantity}
    )
  }
}


module.exports = {Client};
```

This JavaScript client is ready to be used in the latest **single-page application** (**SPA**) frontend and deployed to production. Before we deploy this client, it needs to be tested against the REST API.

Now, instead of starting up the real REST API server and running tests, we want to create individual interactions and test those against a mock provider, then use them to produce a contract that is shared with the provider so that it may verify every interaction from its point of view. We will be able to test these interactions just as swiftly as our unit tests.

To better explain these interactions, we will look at one from `/baskets/ui/client.spec.js` for the UI consumer tests in the **Shopping Baskets** module:

```
provider.given('a store exists')
  .given('a product exists', {id: productId})
  .given('a basket exists', {id: basketId})
  .uponReceiving(
      'a request to add a product with a negative quantity'
  )
  .withRequest({
    method: 'PUT',
    path: `/api/baskets/${basketId}/addItem`,
    body: {
      productId: productId,
      quantity: -1,
```

```
    },
    headers: {Accept: 'application/json'},
  })
  .willRespondWith({
    body: MatchersV3.like({
      message: 'the item quantity cannot be negative',
    }),
    headers: {'Content-Type': 'application/json'},
    status: 400,
  });
```

In the previous listing, we are building an interaction for a call to the AddItem endpoint. We expect to receive an error when we include a negative quantity in our request.

Here is what each method is doing when building the interaction:

- given() is used to signal to the provider that a certain state should be configured or used to respond to the simulated request when it is verifying the contract. Of the four methods shown, only given() is optional. It is used in the code example three times, with two of the calls including static data that should be used in place of the state the provider would generate.

- uponReceiving() sets up a unique name for this expectation.

- withRequest() defines the exact request that will be used by both the consumer tests and provider verification tests. In the consumer tests, it is compared with the real request that will be made to the mock provider. Then, in the provider tests, it will be used as a simulated request from the mock consumer against the provider.

- willRespondWith() is the expected response. We build it using *matchers*, creating an expectation based on what is important to the consumer. In the consumer tests, this response will be returned by the mock provider and in the provider tests, the real response is verified against it. The real error response from the AddItem endpoint includes more than the message property, but we match only the one value that we care about.

The interaction is then tested using your preferred testing library. We will only be able to truly test one side of the interaction right now, which involves verifying that the request we send to the mock provider is exactly as we said it would be:

```
it('should return an error message', () => {
  return provider.executeTest((mockServer) => {
    const client = new Client(mockServer.url);
    return client.addItem(basketId, productId, -1)
      .catch(({response}) => {
```

```
        expect(response.status).to.eq(400);
    });
  });
});
```

To test the interaction with the consumer, we use the real client code to create and send a request to the mock provider. The response can be checked as well, and in this case, we catch the expected error response. If we don't, then an uncaught exception could occur, and it will throw off our test.

When all of our consumer tests are passing, a contract will be generated using the consumer and provider names, such as `baskets-ui-baskets-api.json`. This contract will need to be shared with the provider somehow so that the other half of the tests can take place. Contracts can be shared via the filesystem, by hosting them, or they can be published to Pact Broker.

To verify a contract with a provider, we need to receive simulated requests. However, we need to return real responses from a real provider. This means that we need to stand up just enough of the provider so that real responses can be built and returned to the mock consumer. The provider tests are located in the `/baskets/internal/rest/gateway_contract_test.go` file.

For the **Shopping Baskets** module, we can start up the gRPC and HTTP servers, use test doubles for all of the application dependencies, and still be able to generate real responses. This provider will need to be running in the background so that the mock consumer can send the interactions that each consumer contract has defined.

When performing the verifications for simple APIs, we could start up the provider, configure the verifier, feed in contracts, and be done with our test:

```
verifier.VerifyProvider(t, provider.VerifyRequest{
    Provider:                  "baskets-api",
    ProviderBaseURL:           "http://127.0.0.1:9090",
    ProviderVersion:           "1.0.0",
    BrokerURL:                 "http://127.0.0.1:9292",
    BrokerUsername:            "pactuser",
    BrokerPassword:            "***",
    PublishVerificationResults: true,
})
```

The configured verifier in the prior listing will connect the mock consumer to the provider running on port `9090`, then look for contracts published to our Pact Broker that belong to the `baskets-api` provider. If every interaction is verified for a contract, then we publish that success back to Pact Broker.

However, if any consumers have made interactions that make use of the provider state, as we did in our `baskets-ui` consumer using `given()`, then those states need to be supported; otherwise, the interactions cannot be verified.

For example, to verify the `AddItem` endpoint, we will need to populate the test doubles with a basket, product, and store records. Using provider states will require communication and collaboration between teams. Documentation could be written that lists the state options that the provider supports. Failing these verification tests could block a provider from deploying, so the use of new provider states should be communicated and documented in all cases.

Provider states may optionally accept parameters that allow consumers to customize the interactions that they send and expect to receive back. The following state is used by the consumer:

```
given('a basket exists', {id: basketId})
```

This is supported by the provider with the following:

```go
// ... inside provider.VerifyRequest{}
StateHandlers: map[string]models.StateHandler{
    "a basket exists": func(_ bool, s models.ProviderState)
        (models.ProviderStateResponse, error) {
        b := domain.NewBasket("basket-id")
        if v, exists := s.Parameters["id"]; exists {
            b = domain.NewBasket(v.(string))
        }
        b.Items = map[string]domain.Item{}
        b.CustomerID = "customer-id"
        if v, exists := s.Parameters["custId"]; exists {
            b.CustomerID = v.(string)
        }
        b.Status = domain.BasketIsOpen
        if v, exists := s.Parameters["status"]; exists {
            b.Status = domain.BasketStatus(v.(string))
        }
        baskets.Reset(b)
        return nil, nil
    },
},
```

Supporting this expected state, as well as the ones for products and stores, should be enough to verify the provider for the current UI consumer.

When the `AddItem` endpoint is verified against the interaction with the negative quantity value, it will produce the following result:

```
a request to add a product with a negative quantity
   Given a store exists
   And a product exists
   And a basket exists
   returns a response which
     has status code 400 (OK)
     includes headers
       "Content-Type" with value "application/json" (OK)
     has a matching body (OK)
```

This result comes from the simulated request being sent to our real provider, which responded exactly how it would under normal conditions. The real response was then compared with the expected response, and it all passed.

With that, we have tested both a real request and a real response and have confirmed that they will work both as intended and expected. The REST API will work for every consumer that has created a contract, giving the provider confidence that it can be deployed without it breaking any consumers.

## Message consumer and provider example

Contracts can also be developed by the consumers of asynchronous messages. We will want to expect messages from the consumers and verify that the providers will send the right messages. With asynchronous messaging, there will be no request portion to the test but only an incoming message to process. Likewise, for the provider, we will not receive any request for a message, so the testing pattern changes slightly.

We will create tests for the messages that the **Store Management** module publishes, and test message consumption in both the **Shopping Baskets** and **Depot** modules.

The consumer tests are located in the `/baskets/internal/handlers/integration_event_contract_test.go` and `/depot/internal/handlers/integration_event_contract_test.go` files. These two modules receive messages from the **Store Management** module, which we will discuss later.

For each message that a consumer expects to receive, we must create an expected message entry in our contract with the following code:

```
message := pact.AddAsynchronousMessage()
for _, given := range tc.given {
    message = message.GivenWithParameter(given)
}
assert.NoError(t, message.
    ExpectsToReceive(name).
    WithMetadata(tc.metadata).
    WithJSONContent(tc.content).
    AsType(&rawEvent{}).
    ConsumedBy(msgConsumerFn).
    Verify(t),
)
```

The GivenWithParameter() and ExpectsToReceive() methods should be familiar to you if you read through the REST example.

WithJSONContent() is one of several methods we can use in Go to provide the expected message to the test. The content that we provide as our expected content is built using matchers. We can also use WithMetadata() to provide expectations for the headers or extra information that is published along with the content. This can be seen in the following example for the test of the StoreCreated event:

```
metadata: map[string]string{
    "subject": storespb.StoreAggregateChannel,
},
content: Map{
    "Name": String(storespb.StoreCreatedEvent),
    "Payload": Like(Map{
        "id":       String("store-id"),
        "name":     String("NewStore"),
        "location": String("NewLocation"),
    }),
}
```

The AsType() method is a convenient way to convert the JSON that results from the matchers into something we can more easily work with and is optional.

Contract testing messaging will not use a mock provider or consumer, which is what we did in the REST example. The consumers will only be receiving messages and are not expected to send anything back. We will not be using a mock provider this time; instead, we will use a function that we provide to ConsumedBy() to test that our expected message will work.

The idea remains the same as in the REST example: we want to test that the message can be consumed. If it cannot, then we need to fix the message, application, or test.

To test that the events we receive work, we will need to turn rawEvent into an actual ddd.Event event, which means also converting the JSON payload into a proto.Message protocol. First, we need to register the storespb.* messages using a JSON Serde instead of the Protobuf Serde we typically use:

```
reg := registry.New()
err := storespb.RegistrationsWithSerde(
    serdes.NewJsonSerde(reg),
)
```

Then, in the function that we provide to the ConsumedBy() method, we will deserialize the JSON into the correct proto.Message:

```
msgConsumerFn := func(contents v4.MessageContents) error {
    event := contents.Content.(*rawEvent)
    data, err := json.Marshal(event.Payload)
    if err != nil { return err }
    payload, err := reg.Deserialize(event.Name, data)
    if err != nil { return err }
    return handlers.HandleEvent(
        context.Background(),
        ddd.NewEvent(event.Name, payload),
    )
}
```

The test will fail if the built event is not handled as expected. For extra measure, we use mocks that are passed into the handlers to test whether the right calls are being made when we call down into the handlers.

The contracts that we produce from message testing will not contain interactions and cannot be verified using a provider test, which is what we used in the REST example. The providers will use the description and any provider states to construct the message that is expected by consumers. There will not be any requests coming in.

Like the **Shopping Basket** REST provider, we want to avoid manually generating the message and should stand up enough of the module to create messages for us. We should verify that the processes that produce messages will continue to produce the right messages into the right streams as the application changes.

Just as we did in the REST provider test, we will create a verifier that will connect to Pact Broker, fetch the contracts that belong to the provider, verify the messages, then publish the results of the verifications back to Pact Broker.

The **Store Management** module provider verification tests can be found in the `/stores/internal/handlers/domain_events_contract_test.go` file. The key differences between this test file and the one for the REST contracts are that we do not start any mock consumer or start the provider listening on any ports. Message verification will also require that we implement each description string that the consumers have used in their contracts, such as "*a StoreCreated message*," as a message handler:

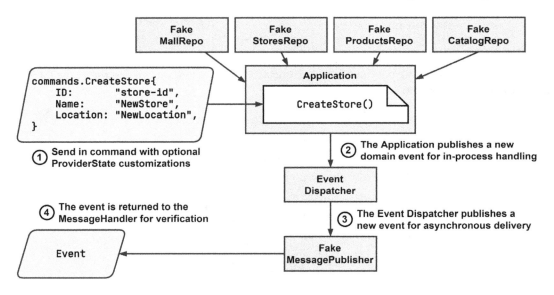

Figure 10.10 – Verifying the StoreCreated message

To verify the `StoreCreated` message, we can make a call into `CreateStore()` that will fire off the domain event, which, in turn, publishes the expected message. Using a `FakeMessagePublisher` test double, we can retrieve the last published message to complete the verification process.

The message payload, `proto.Message`, is serialized using a JSON Serde, similar to what we used in the consumer tests. We need to use the same methods for encoding when we create these messages tests, and JSON is currently the best option for the content that we want to verify. Other formats could be used but the Pact tools support JSON the best and the matchers only work with JSON.

Our entire message handler for the `StoreCreated` event message looks like this:

```
"a StoreCreated message": func(
      states []models.ProviderState,
  ) (message.Body, message.Metadata, error) {
  // Arrange
  dispatcher := ddd.NewEventDispatcher[ddd.Event]()
  app := application.New(
      stores, products, catalog, mall, dispatcher,
  )
  publisher := am.NewFakeMessagePublisher[ddd.Event]()
  handler := NewDomainEventHandlers(publisher)
  RegisterDomainEventHandlers(dispatcher, handler)
  cmd := commands.CreateStore{
      ID:        "store-id",
      Name:      "NewStore",
      Location: "NewLocation",
  }
  // Act
  err := app.CreateStore(context.Background(), cmd)
  if err != nil { return nil, nil, err }
  // Assert
  subject, event, err := publisher.Last()
  if err != nil { return nil, nil, err }
  return rawEvent{
          Name:     event.EventName(),
          Payload: reg.MustSerialize(
              event.EventName(), event.Payload(),
          ),
      }, map[string]any{
          "subject": subject,
      }, nil
},
```

The real, albeit `rawEvent` event, is returned, along with a map for the metadata containing the subject that the message, if it had been published, would have been published into.

With that, we have completed the message verification process. We have taken a contract containing the expected messages for a pair of consumers and verified them with the provider. The results are automatically published to Pact Broker. If configured, Pact Broker could then inform the CI/CD processes to allow deployments to proceed.

Contract testing allows us to test integrations between components very quickly and with a lot less effort than if we had used a more traditional integration test approach. We can test the integration between two components, but we still need to test the operations that span multiple interactions.

## Testing the application with end-to-end tests

The final form of testing we will cover is **end-to-end** (**E2E**) testing. E2E testing will encompass the entire application, including third-party services, and have nothing replaced with any test doubles. The tests should cover all of the processes in the application, which could result in very large tests that take a long time to complete:

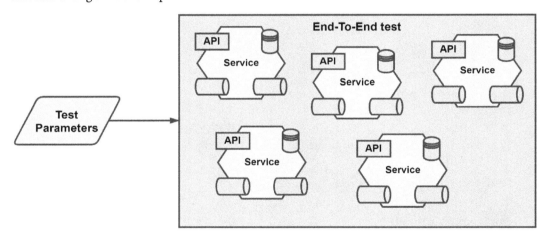

Figure 10.11 – The scope of an end-to-end test

E2E testing takes many forms, and the one we will be using is a features-based approach. We will use Gherkin, introduced in *Chapter 3*, *Design and Planning*, to write plain text scenarios that should cover all essential flows throughout the application.

## Relationship with behavior-driven development

You can do **behavior-driven development** (**BDD**) without also doing E2E testing, and vice versa. These two are sometimes confused with each other or it's thought that they are the same. BDD, as a practice, can be used at all levels of the testing pyramid and not just for the final E2E tests or the acceptance tests. Whether or not to also employ BDD, and perhaps TDD, is a tangential decision for any particular level of testing in your testing strategy:

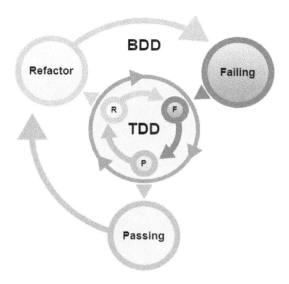

Figure 10.12 – The double-testing loop with BDD and TDD

BDD is also associated with the Gherkin language, and it has become dominant because of how the user stories BDD uses are created. We will be using Gherkin to write our features and their related scenarios but again, this does not mean that we will be doing BDD. Gherkin can also be used for our unit or integration tests. Instead of using table-driven tests or a library to run tests as a suite, they could be written as plain text tests.

## E2E test organization

Our first step in E2E testing is to create feature specifications and then record them in our feature test files using Gherkin. There is no standard for organizing these feature files, but if we consider that an application uses multiple repositories because it is a distributed application that also uses microservices, then organizing all of the features into a repository might make sense. We only have one repository, so we will organize all of the features and other E2E-related test files under /testing/e2e.

## Making executable specifications out of our features

To make a feature file an executable specification, we will use the godog library, which is the official Cucumber (https://cucumber.io) library for Go. With this library, we can write a TestEndToEnd function that will be executed using the go test command.

We will also need clients for each of the REST APIs. Normally, E2E tests would involve interacting with some end user UI, but our little application has none to work with at the moment. The REST clients can be generated using the **go-swagger** (https://github.com/go-swagger/go-swagger) tool, which can be installed along with the other tools we have used in this book by running the following command from the root of the code for this chapter:

```
make install-tools
```

The actual command to generate the clients is then added to the generate.go file for each module. The added command looks something like the following, with added line breaks to make it easier to read:

```
//go:generate swagger generate client -q
  -f ./internal/rest/api.swagger.json
  -c storesclient
  -m storesclient/models
  --with-flatten=remove-unused
```

The generate command in the previous listing will create an entire REST client that is ready to be pointed at the **Store Management** REST API.

The final step of turning features into executable specifications is to implement each step and then register the implementation with the library.

### *Example step implementation*

Let's say we have the following feature:

```
Feature: Register Customer

    Scenario: Registering a new customer
        Given no customer named "John Smith" exists
        When I register a new customer as "John Smith"
        Then I expect the request to succeed
        And expect a customer named "John Smith" to exist
```

We have four steps that we need to implement and register. To implement the registration of a new customer, we can start with a function signature, like this:

```
func iRegisterANewCustomerAs(name string)
```

The string that is enclosed within the double quotes would be passed as the name parameter. Steps can have several parameters, and those parameters can be of several different Go types. Gherkin *Docstrings* and *Tables* are supported and can be passed in as well. The name of the function does not matter to the library and can be anything.

The function can be standalone or be part of a struct if you want to capture and use some test state, for example. We can also have an error return value if the step should fail:

```
func iRegisterANewCustomerAs(name string) error
```

After we have implemented our step, we will need to register it so that when godog runs across the step statement, it knows what function will be expected to handle it:

```
// ctx is a *godog.ScenarioContext
ctx.Step(
    `^I register a new customer as "([^"]*)"$`,
    iRegisterANewCustomerAs,
)
```

The step statements may be provided as strings and are interpreted as regular expressions, or directly as a compiled *regexp.Regexp. This is so that the parameters can be parsed out and passed into the step function.

## What to test or not test

E2E testing sits very high on the testing pyramid, and we should not try to write features covering everything that the application does or can do. Start with the critical flows to the business and then go from there. The identified flows will have several tests associated with them, not just one. You will want to consider what conditions can affect it and write tests to cover those conditions.

Some flows may not automate very well and should be left for the testers to run through manually.

# Summary

Testing an event-driven application is no harder than testing a monolithic application when you have a good testing strategy. In this chapter, we covered the application, domain, and business logic using unit tests. These tests make up the bulk of our testing force. We follow up our unit tests with integration tests, which help uncover issues with how our components interact. Using tools such as Testcontainers-Go can help reduce the effort required to run the tests, and using libraries such as the Testify suite can help reduce the test setup and teardown complexities.

A distributed application, whether it is event-driven like ours or synchronous, gains a lot from including contract testing in the testing strategy. Having confidence in how you are using or have made expectations of a provider without the mess and complexities of standing the provider up is a time saver many times over. Finally, including E2E testing in any form will give the team and stakeholders confidence that the application is working as intended.

In the next chapter, we will cover deploying the application into a Kubernetes environment. We will be using Terraform so that our application can be deployed to any cloud provider that provides Kubernetes services. We will also break a module out of the monolith into a microservice so that we can deploy it.

# 11

# Deploying Applications to the Cloud

In this book, we have worked with the MallBots application as a modular monolith and have only experienced running it locally using `docker compose`. In this chapter, we will be breaking the application into microservices. We will update the Docker Compose file so that we can run either the monolith or the microservices. Then, we will use **Terraform**, an **Infrastructure as Code (IaC)** tool, to stand up an environment in AWS and deploy the application there.

In this chapter, we are going to cover the following topics:

- Turning the modular monolith into microservices
- Installing the necessary DevOps tools
- Using Terraform to configure an AWS environment
- Deploying the application to AWS with Terraform

# Technical requirements

You will need to install or have installed the following software to run the application or to try the examples:

- The Go programming language **version 1.18+**
- Docker
- The Kubernetes CLI tools
- Terraform
- The AWS CLI
- The PostgreSQL CLI tools

We have a lot more tool requirements for this chapter and will be covering download locations and installation within the chapter for each new tool. The code for this chapter can be found at `https://github.com/PacktPublishing/Event-Driven-Architecture-in-Golang/tree/main/Chapter11`.

# Turning the modular monolith into microservices

Our application, while it is modular, is a monolith. It is built as a single executable and can be deployed as a single application. There is nothing wrong with that but faced with scaling issues, we have only one knob we can adjust. If we broke the application up by turning each module into its own microservice, then when faced with scaling issues, we would have finer control over how the application can be deployed to support the load.

Turning our application into microservices will have many steps to it but will not be difficult:

1. We will need to refactor the monolith construct used to initialize each module.
2. We will make some small updates to the composition root of each module.
3. We will then update each module so it can run as standalone.

After we are done with these steps, we will update the Docker Compose file and make other small changes so that the two experiences, running the monolith or running the microservices, are the same.

## Refactoring the monolith construct

Our motivation for updating this part of the application is so that we can continue to run the monolith after we have turned each module into a microservice:

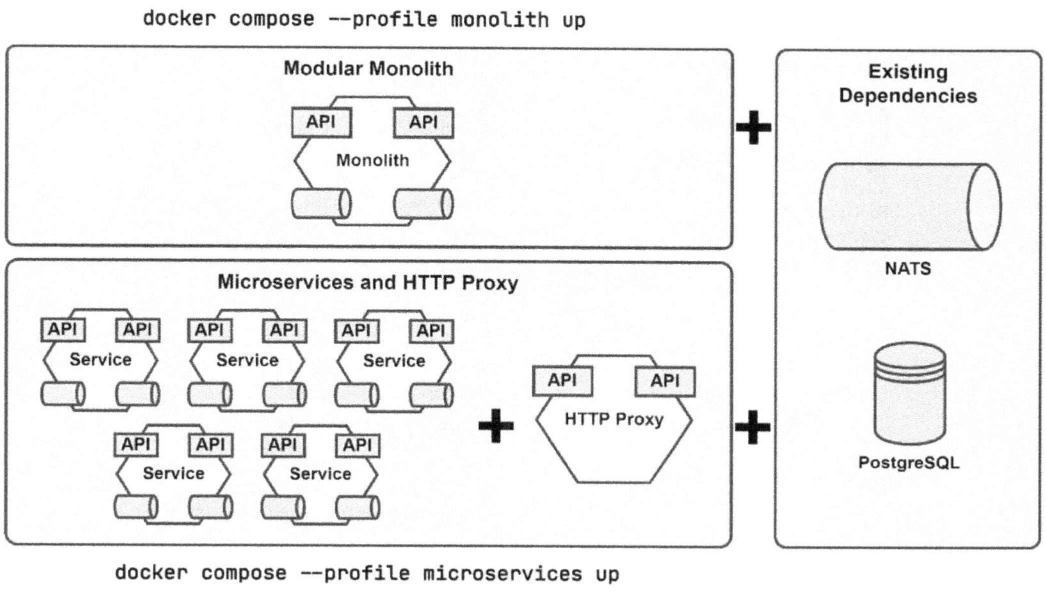

docker compose --profile monolith up

docker compose --profile microservices up

Figure 11.1 – Docker Compose with either a monolith or microservices

The monolith is built using the /cmd/mallbots main package. Up to this point, we have used a local app struct in that package to provide each module the resources that they required. The unexported app struct implements the Monolith interface and this interface was used in each module's Startup() method.

> **Docker Compose version**
>
> The Docker Compose command, docker compose, that I am using is available from the **Compose V2** release. If this command is not available, you can use the older version by putting a hyphen between the words as follows: docker-compose. The arguments used in the examples will not change when using the older version of the command.

Using the app struct as a template, we will create a new shared System struct in a new /internal/ system directory and package:

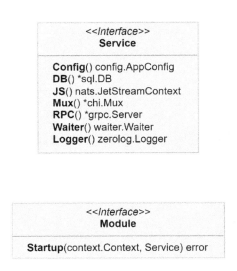

Figure 11.2 – The types and interfaces of the system package

This new package also duplicates the interfaces that were found in the `monolith` package. From the `monolith` package, the old `Monolith` interface is renamed `Service` because it will serve a new more general service need.

We can see from *Figure 11.2* that the `System` struct has a lot of the same functionality, with some new exported methods, as the original `app` struct that it will be replacing. We did not bring over anything to do with managing the modules. Working with the modules will remain a monolith-only concern and we use the following to reimplement the existing functionality for `/cmd/mallbots`:

```
type monolith struct {
    *system.System
    modules []system.Module
}
func (m *monolith) startupModules() error {
    for _, module := range m.modules {
        ctx := m.Waiter().Context()
        err := module.Startup(ctx, m)
        if err != nil { return err }
    }
    return nil
}
```

The original `app` was initialized using some small functions. This initialization can be moved into functions alongside or as methods on `System`. These can all be called from a new constructor for `System` such as the following:

```
func NewSystem(cfg config.AppConfig) (*System, error) {
    s := &System{cfg: cfg}
    if err := s.initDB(); err != nil { return nil, err }
    if err := s.initJS(); err != nil { return nil, err }
    s.initMux()
    s.initRpc()
    s.initWaiter()
    s.initLogger()
    return s, nil
}
```

The original monolith `main.go` file can now be switched over to use `System` instead of `app` and the unused functions and the `monolith.go` file are removed. The `internal/monolith` directory can also be removed. Any lingering references to its package will be addressed in the upcoming section when we turn our attention to the modules.

## Updating the composition root of each module

Every module uses a `Startup()` method to initialize itself to run with the resources that the monolith has provided. Our update will be a small one. We will be moving the code within `Startup()` to a new `Root()` function. Then, we create a call to it from `Startup()` and it will be as though nothing has changed:

```
func (m *Module) Startup(
    ctx context.Context, mono system.Service,
) (err error) {
    return Root(ctx, mono)
}
func Root(
    ctx context.Context, svc system.Service,
) (err error) {
    // ...
}
```

This simple change will allow us to reuse the composition root code for the other method of running the module, running it as a standalone microservice. We do not need to maintain the composition root in this way if we do not want to continue to run the monolith. If a real-world application were to be organized this way and the opportunity presented itself, why not keep the option to run a single process? Being able to continue to run the application as a monolith would allow us to avoid some of the trade-offs with a microservices architecture. For example, local development becomes more resource intensive because more resources will be required to have each service running. Also, attaching a debugger to debug a single process is much easier than attaching multiple debuggers to multiple processes.

## Making each module run as a service

Each module will be made to run standalone by adding /<module>/cmd/service and a new main package to them. These additions are little more than copies of the monolith version. We remove anything to do with the management of modules and are left with the following:

```go
func main() {
    if err := run(); err != nil {
        fmt.Printf(
            "baskets exited abnormally: %s\n", err,
        )
        os.Exit(1)
    }
}
func run() (err error) {
    var cfg config.AppConfig
    cfg, err = config.InitConfig()
    if err != nil { return err }
    s, err := system.NewSystem(cfg)
    if err != nil { return err }
    defer func(db *sql.DB) {
        if err = db.Close(); err != nil { return }
    }(s.DB())
    err = s.MigrateDB(migrations.FS)
    if err != nil { return err }
    s.Mux().Mount("/",
        http.FileServer(http.FS(web.WebUI)),
    )
    err = baskets.Root(s.Waiter().Context(), s)
```

```
    if err != nil { return err }
    fmt.Println("started baskets service")
    defer fmt.Println("stopped baskets service")
    s.Waiter().Add(
        s.WaitForWeb,
        s.WaitForRPC,
        s.WaitForStream,
    )
    return s.Waiter().Wait()
}
```

We replaced the setup of the modules with a single call to this module's Root() function.

Thanks to moving the bulk of the initialization of the system to the constructor, starting up the monolith or each service does not take much. Again, we must consider what trade-offs were made by refactoring things this way. If the microservices begin to diverge in the resources that they need, then we may end up initializing resources for dependencies that we do not have. System is a simple construct that starts up everything the same way – when the need arrives, it can be updated to be smarter about what should be initialized and what should not.

Every module could be run standalone at this point, but we would run into a few issues if we tried to copy the monolith service for each new service into the docker-compose.yml file.

Running our services and having the same experience as running the monolith will require a few more updates to be made.

## Updates to the Dockerfile build processes

We have only a single Dockerfile that builds the monolith. Going forward, we also need a way to compile the individual services. To accomplish this, I will use an additional Dockerfile that will make use of build arguments to target the right service to build.

The new Dockerfile will be named Dockerfile.microservices and live alongside the current one in /docker:

```
ARG svc

FROM golang:1.18-alpine AS builder
ARG svc
WORKDIR /mallbots
COPY go.* ./
RUN go mod download
```

```
COPY .. ./
RUN go build -ldflags="-s -w" -v -o service \
    ./${svc}/cmd/service

FROM alpine:3 AS runtime
COPY --from=builder /mallbots/docker/wait-for .
RUN chmod +x /wait-for
COPY --from=builder /mallbots/service /mallbots/service
CMD ["/mallbots/service"]
```

This is a multi-stage `Dockerfile`. In our first stage called `builder`, we compile the service into a binary. In the second stage, we copy the `wait-for` utility, which is used to wait for the database to be available, and the newly compiled binary. By using this `Dockerfile`, we keep the containers we produce very small, which helps with transferring them and loading them, among other things.

To build the specific service, we want we use the `--build-arg=svc=<service>` command-line argument with the `docker build` command as follows:

```
docker build -t baskets --file
docker/Dockerfile.microservices --build-arg=svc=baskets .
```

This command would build the **Shopping Baskets** microservice and make it available as baskets in our Docker repository.

## Updates to the Docker Compose file

We will need to update the `docker-compose.yml` file so that each microservice can be started much like the monolith was. First, we need to add in each service using a block of YAML such as the following:

```
baskets:
  container_name: baskets
  hostname: baskets
  image: baskets
  build:
    context: .
    dockerfile: docker/Dockerfile.microservices
    args:
      service: baskets
  ports:
```

```
    - '8080:8080'
expose:
    - '9000'
environment:
    ENVIRONMENT: development
    PG_CONN: <DB CONNECTION PARAMS>
    NATS_URL: nats:4222
depends_on:
    - nats
    - postgres
command:
    - "./wait-for"
    - "postgres:5432"
    - "--"
    - "/mallbots/service"
```

Similar blocks are added for the other modules-turned-microservices that we want to start up (in total, nine new blocks of YAML are added). Secondly, we want to be able to start either the monolith or the microservices version of our application. To do that, we can use the `profiles` feature of Docker Compose to selectively start services. At the end of the monolith `services` block, we append the following YAML:

```
services:
  monolith:
    # ... existing YAML
    profiles:
      - monolith
```

We can do the same for each new service, except using microservices instead:

```
services:
  baskets:
    # ... existing YAML
    profiles:
      - microservices
```

With those last edits made to the `docker-compose.yml` file, we can start the monolith or start the microservices version of our application.

### Starting the monolith

Running the following command will run NATS, PostgreSQL, the Pact Broker, and then only the monolith service:

```
docker compose --profile monolith up
```

It is the same experience we are used to running, only that now we also need to include the `–profile monolith` part to get it.

### Starting the microservices

Running the following command will appear the same at first, with a lot more containers to build:

```
docker compose -profile microservices up
```

However, it will fail to run, ending with the following error message:

```
Bind for 0.0.0.0:8080 failed: port is already allocated
```

This is because we have configured each microservice to use the same host port for their HTTP port. We can fix this by changing the host port each microservice uses:

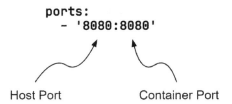

Figure 11.3 – Host and container ports for docker compose services

To give each microservice a unique but memorable new host port, we will use a sequence starting with the `baskets` entry down to the `stores` entry. For the `baskets` entry, we will use port `8081` and for `stores`, we will be using `8089` for its host port. The container port should remain as it is for all microservices.

Running the `compose` command again starts up the environment. Requests need to be sent to the correct port for the service now. If we attempt to open the **Swagger UI**, we will run into our second problem. We cannot load the OpenAPI specifications as we could before when we were running the monolith. The local specification for each service can be loaded, but we will not be able to view them all as we could before:

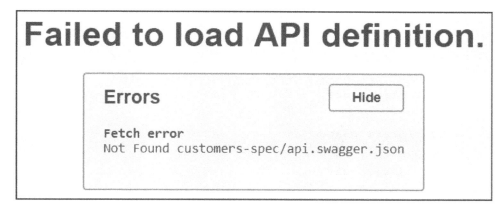

Figure 11.4 – The Swagger UI experience is broken

Our fix has allowed our microservices to run but the overall experience is far from the same as it was with the monolith. To fix this current problem with loading the OpenAPI specifications and to return the experience to what it was before, we will need to add a reverse proxy service to the `docker-compose.yml` file.

A reverse proxy will take the requests we send in and direct them to one of our microservices. The client will only interact with the reverse proxy and will not be aware of the microservices behind it.

## Adding a reverse proxy to the compose environment

We can quickly set up a reverse proxy using **Nginx**. Nginx is a popular web server, reverse proxy, and load balancer application. We only need to set up a reverse proxy today and thankfully, it is going to be quite easy to do.

First, we define a configuration file for the application called `/docker/nginx.conf`:

```
worker_processes 1;
events { worker_connections 1024; }
http {
  sendfile on;

  upstream docker-baskets {
    server baskets:8080;
  }
# ... plus upstream blocks for each other microservice

  server {
```

```
    listen 8080;

    location /api/baskets {
      proxy_pass      http://docker-baskets;
      proxy_redirect off;
    }
    location /baskets-spec/ {
      proxy_pass      http://docker-baskets;
      proxy_redirect off;
    }
# ... plus location block pairs for each other microservice
# then one more location block for the swagger-ui files
    location / {
      proxy_pass      http://docker-baskets;
      proxy_redirect off;
    }
}
```

I have only included the `baskets` microservice as an example in the configuration file example, but each microservice would need to have an upstream configuration and a pair of location configurations so that the reverse proxy could properly redirect the requests to where they need to go. A final location block is used to serve the **Swagger UI** from any microservice.

Secondly, we need to remove the port configurations for each microservice and add the reverse proxy as a new service to the `docker-compose.yml` file:

```
reverse-proxy:
  container_name: proxy
  hostname: proxy
  image: nginx:alpine
  ports:
    - '8080:8080'
  volumes:
    - './docker/nginx.conf:/etc/nginx/nginx.conf'
  profiles:
    - microservices
```

We only want the reverse proxy to start up with the other microservices, so it is also given the `microservices` profile. The configuration file we created for Nginx is mounted at the appropriate place for the application to find it.

We have now fixed the initial experience and can use the Swagger UI again, but we will quickly run into our final problem if we run the E2E tests. The services that use gRPC fallbacks are still dialing into gRPC connections that point back to the local gRPC connection. When the **Shopping Baskets** service tries to make a fallback call to the **Store Management** service to locate a product, it calls itself. We will need to provide the addresses of the services somehow so that proper connections can be made.

## Fixing the gRPC connections

We will provide the addresses of the gRPC servers that are being used by other services with a new environment variable called `RPC_SERVICES`. This value will be a map of the service names and their addresses:

```
RPC_SERVICES="STORES=stores:9000,CUSTOMERS=customers:9000"
```

In the `/internal/rpc/config.go` file, we add the following:

```
type RpcConfig struct {
    // ... snipped existing fields
    Services
}
type Services map[string]string
func (c RpcConfig) Service(service string) string {
    if address, ok := c.Services[service]; ok {
        return address
    }
    return c.Address()
}
func (s *Services) Decode(v string) error {
    services := map[string]string{}
    pairs := strings.Split(v, ",")
    for _, pair := range pairs {
        p := strings.TrimSpace(pair)
        if len(p) == 0 {
            continue
        }
        kv := strings.Split(p, "=")
```

```
        if len(kv) != 2 {
                return fmt.Errorf("invalid pair: %q", p)
        }
        services[strings.ToUpper(kv[0])] = kv[1]
    }
    *s = services
    return nil
}
```

The `Services` type will use the custom decoder, `Decode()`, to turn the service pairs into a usable map. A `Service()` method is also added to the `RpcConfig` struct for convenience so it will be easier to fetch the correct service address when we need to.

Now, we need to update the **Shopping Baskets** and **Notifications** composition roots to use the correct address for the gRPC connection that they are dialing into. Here is the updated connection from **Shopping Baskets** to the **Store Management** service:

```
container.AddSingleton("storesConn",
    func(c di.Container) (any, error) {
        return grpc.Dial(
            ctx,
            svc.Config().Rpc.Service("STORES"),
        )
    },
)
```

Now, all that is left to do is to add the new `RPC_SERVICES` environment variable to each service in the `docker-compose.yml` file:

```
# ... snipped other configuration
environment:
  # ... snipped other variables
  RPC_SERVICES: 'STORES=stores:9000
      ,CUSTOMERS=customers:9000'
```

Rebuild the microservice containers and restart the `compose` environment and now our E2E tests all pass again. Likewise, trying to add an item to a basket with an invalid product identifier in the Swagger UI also behaves as expected, if you care to verify things are working that way.

Our application can now run as a monolith or as a suite of microservices. To recap, these are the steps we took to get here:

1. We refactored the monolith start up code into a shared service start up library.

2. Each module got a new service command, with an updated composition root.

3. To build the new services, a new `Dockerfile` was created that used build arguments so that a single `Dockerfile` could be used for all services.

4. The `docker-compose.yml` file was updated to include each service, and we used `docker compose` profiles to start either the monolith or the microservices.

5. A reverse proxy was added so we could reach all services with a single address.

6. We updated the gRPC configuration so we could provide the right gRPC server addresses to the gRPC clients.

We will also want to run our application in the cloud, and we have many providers to choose from. **Amazon Web Services (AWS)**, at `https://aws.amazon.com`, is the oldest, largest, and most well-known cloud provider. There are other big names to choose from, such as **Google Cloud Platform (GCP)**, at `https://cloud.google.com`, and **Azure Cloud**, at `https://azure.microsoft.com`. Smaller or regional cloud providers are also available, such as **Digital Ocean** (`https://www.digitalocean.com`), **OVHcloud** (`https://ovhcloud.com`), and **Hetzner** (`https://www.hetzner.com`).

From all of these options, we will be using AWS, partly because of its status as the top cloud provider and partly because it is also the one I know best. However, before we do that, we will need to install and get a little familiar with some new tools.

# Installing the necessary DevOps tools

The plan is to deploy the application in its microservices form to AWS. For most developers, learning about every service offering in AWS is not something they focus on – taking off their software developer hat and putting on their system administrator hat, so to speak. To make things easier, we will be relying on an application called **Terraform**, which is an **IaC** tool. We will be able to define what our application needs with code and then let it do the heavy lifting of pulling all the right levers and pushing all the right buttons for us.

We will also need a few more tools to help us:

- The AWS CLI, `aws`, is how we will authorize ourselves with AWS

- Helm is a tool that will let us use packages called Charts to deploy some complex machinery into Kubernetes

- We will be using a **PostgreSQL** database in the cloud and will want the **PostgreSQL client psql** installed to help set it up

- To view our Kubernetes cluster, we will use an application called **K9s**, which is a **Terminal UI (TUI)** that makes it very easy to navigate around the cluster
- We will also need a tool called **Make**, which is a small application runner that helps us turn large or multistep commands into ones that are easy to remember and run

If you do not already have these applications installed, I have two options for you to install them. The first option is to keep your local system clean of additional applications by using a Docker container with all of the applications already installed or to find and install them yourself.

If you are going to be following along and you are on Windows, I recommend the first option.

Regardless of which option you choose, you will also need an AWS account. Visit `https://portal.aws.amazon.com/billing/signup` to create a free account with AWS. Let us check both options.

## Installing every tool into a Docker container

This is the easier route to take and it also keeps your local system clean of any applications you are not likely to be using again. This option will compile a Docker container called `deploytools`, which will then be made available with a shell command alias.

To start you need to either be using macOS or Linux or be able to open a Powershell in Windows. A non-Powershell Command Prompt in Windows will not work.

To start, go into `deployment/setup-tools` in your Terminal or Powershell window.

You will now need to execute the right script for your OS. macOS and Linux users should run the following command:

```
source set-tool-alias.sh
```

Powershell users should run this command:

```
.\win-set-tool-alias.ps1
```

Both do the same things. During the first run, the `deploytools` container will be built; subsequent runs will rebuild the container only if it is missing or the `Dockerfile` has changed. It will then set up the `deploytools` command. This is a temporary command that will stop working when you close the Terminal or window. To get it back, you just need to run the correct script command again from the `deployment/setup-tools` directory.

Once you have your alias, you can verify it works by running the following:

```
deploytools terraform -version
```

You should see the Terraform version printed out, looking something like this:

```
Terraform v1.2.9
on linux_amd64
```

If you see that, then the container and command are ready for use.

When you are using this option, you need to prefix the commands in the following sections with the `deploytools` command. Let's take this command as an example:

```
aws configure
```

Turn it into this command:

```
deploytools aws configure
```

Speaking of which, you will still need to configure your AWS credentials. You will find instructions to do so in the *Creating and configuring your AWS credentials* section that comes a little later.

Next, let's look at the other option.

## Installing the tools into your local system

We will need a few tools to support our plans to deploy the application as microservices in AWS. All of these tools are available for Linux, macOS, and Windows OSs; only the download location or installer will be different. Using them will be the same.

### Installing and configuring the AWS CLI

The first tool we will want to install is the AWS CLI. You can find instructions for your OS at `https://docs.aws.amazon.com/cli/latest/userguide/getting-started-install.html`.

Once you have downloaded the tool, we need to set up a user and configure it in your shell.

### Creating and configuring your AWS credentials

We will be using this user to access the AWS services from Terraform and the CLI:

1. Sign into your AWS account.
2. Access the **Identity and Access Management (IAM)** service.
3. Click on **Users** on the sidebar.
4. Click on the **Add User** button to begin creating your user.
5. Give the user a name, such as `mallbots_user`, choose **Access Key** as the credential type, and then click on the **Next** button.

6.  Choose to attach existing policies directly, select the **AdministratorAccess** permission, and then click on the **Next** button.

7.  You may add any tags you wish – then, click on the **Next** button.

8.  Confirm that the user has *Programmatic access* and is using the **AdministratorAccess** policy and then click on the **Create user** button.

On this next screen, you should download the credentials as a .csv file. Do not close this page before getting the credentials, as this is the only time you will be given the opportunity to retrieve them.

When you are done testing the deployment of the application and have properly removed all of the resources, you can then remove this user. There are no charges incurred by keeping this user account present.

Next, we will use aws-cli to configure your shell with the credentials you just downloaded. Locate and open the .csv file with your credentials to use them with the following command:

```
aws configure
```

You will be prompted to enter the user's access key ID, secret access key, default region, and default output format. The keys are found in the .csv file you downloaded. You may leave the default values blank if you wish. For the default region, there are many options to select from and you should select the region that is nearest to you.

To verify that you have entered the credentials correctly, use the following command to fetch a list of users from IAM:

```
aws iam list-users
```

If you see a list of users, including the user you just made, then aws-cli is ready to use.

### Installing Terraform

The installers for Terraform can be found at https://learn.hashicorp.com/tutorials/terraform/install-cli. **Version 1.2.8** was used for the examples in this book.

Once Terraform has been installed for your operating system, it requires no configuration and is ready to use.

### Installing Helm

Some of the configurations we will be using will be in the form of Helm charts, which are collections of files that describe Kubernetes resources. Instead of creating new custom Terraform code, we can rely on the battle-tested community versions.

Helm install instructions can be found at `https://helm.sh/docs/intro/install/`. **Version 3.9.4** was used for the examples in this book.

### Installing the tools to access Kubernetes clusters

We will be deploying the application into a Kubernetes cluster and instead of navigating the AWS console to keep an eye on things, we will install some tools to make things easier.

The first tool we will install is **K9s**, a TUI application that makes it very easy to browse the various resources, such as Pods, Ingresses, and Services that will be part of the Kubernetes cluster. The install instructions can be found at `https://k9scli.io/topics/install/`. **Version 0.26.3** was used for the examples in this book.

The second optional tool to install is **kubectl** and the installers can be found at `https://kubernetes.io/docs/tasks/tools/`. **Version 1.25.0** was used for the examples in this book.

Other tools will need a Kubernetes configuration before they can be used and we will be able to fetch one after we deploy the infrastructure to AWS.

### Installing the tools to initialize PostgreSQL

We will be using the PostgreSQL CLI tool psql to initialize the databases and set up the schemas and users after we deploy the application infrastructure. The psql tool comes with the PostgreSQL server installation. We do not need to install the PostgreSQL server, so if you are given the option, choose to only install the command-line tools. The PostgreSQL installers can be found at `https://www.postgresql.org/download/`. Not every installer will put the psql tool in your path; you will have to either move the file or add the install location to your path. **Version 14.5** was used for the examples in this book.

We now have our environment ready to execute the deployment scripts and configurations to deploy the infrastructure and application up to AWS.

# Using Terraform to configure an AWS environment

The MallBots application is going to be run from **AWS Elastic Kubernetes Service (EKS)**, a managed Kubernetes environment. The IaC to create the infrastructure is going to be found in the `/deployment/infrastructure` directory.

We will be configuring a small typical AWS environment across two **Availability Zones (AZs)**:

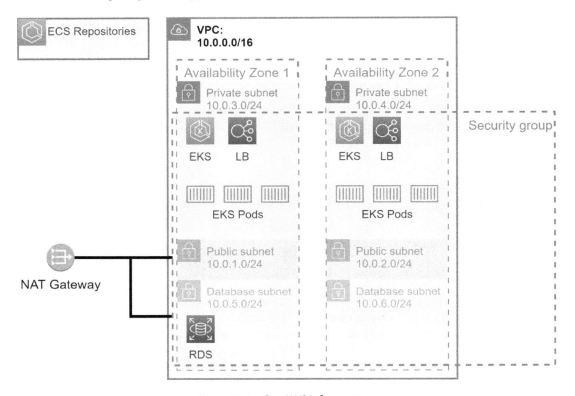

Figure 11.5 – Our AWS infrastructure

In the infrastructure directory, there are several Terraform files. Altogether, they are going to be used to set up the following in AWS:

- Docker repositories with **Elastic Container Service (ECS)**. We will be uploading the built microservice images here.

- A Kubernetes cluster in EKS. We will be deploying our application here from images stored in ECS.

- A PostgreSQL database using **Relational Database Service (RDS)**. A single instance will serve all of the microservice databases and schemas.

- Additional components such as a **Virtual Private Cloud (VPC)** and its subnets. Security groups, roles, and policies to both permit and lock down access.

When running the next commands, you need to be in the /deployment/infrastructure directory.

## Preparing for the deployment

Terraform is capable of deploying thousands of different kinds of resources but it cannot do it by itself. We will need to install the libraries that our specific project needs and to do that, we need to run the following command:

```
make ready
```

This will run both the `terraform init` and `terraform validate` commands. The `init` command will download the libraries and executables needed by the scripts that have been written to build our environment. The `validate` command will also validate our scripts are correct.

The next Terraform command that we run is going to ask for some input from us. Instead of providing the input each time we run it, we can provide the values automatically with a variables file. Create a file named `terraform.tfvars` and put the following lines into it:

```
allowed_cidr_block = "<Your Public IP Address>/32"
db_username = "<Preferred DB username>"
region = "<Your Nearest Or Preferred AWS Region>"
lb_image_repository = "<AWS Regional Image Repository>"
```

The first variable is used to limit access to the resources that are created to your IP or a block of IP addresses. If you only want to allow your public IP, then keep `/32` at the end – for example, `192.168.13.13/32`. The DB username will be used along with a generated password to connect to the database to initialize it in a subsequent step. The final two variables should be set to the AWS Region that works best for you. You can find which repository to enter at `https://docs.aws.amazon.com/eks/latest/userguide/add-ons-images.html`.

It is not critical you create this file but if you do not, then you will be prompted for the values each time you create a new Terraform deployment plan.

## A look at the AWS resources we are deploying

The AWS resources that we will be deploying are broken up into different files, so let's run through each file and cover the major resources that will be installed and configured by the Terraform code within them:

- **Application Load Balancer (ALB):** The `alb.tf` file sets up a service account on the Kubernetes cluster that will be used by the ALB. The file also contains a Helm resource that will install the ALB using a Chart.

- **Elastic Container Registry (ECR):** The `ecr.tf` file sets up private image registries for each of the nine services we will be deploying. It will also build and push each service up into the newly created registries.

- **EKS**: The `eks.tf` file is responsible for creating our Kubernetes cluster. It makes use of a Terraform module, which is a collection of other Terraform scripts, to build the necessary resources from one resource definition. Some AWS IAM policies and roles are configured in this file for the cluster to support the installation of the ALB.

- **RDS**: `rds.tf` will set up a serverless PostgreSQL database and make it available to the Kubernetes cluster. The database will also be accessible by us or anyone else who has an IP address allowed by the `allowed_cidr_block` value.

- **Security groups**: The `security_groups.tf` file will set up our security group that will limit access to our resources from the internet. Whatever `allowed_cidr_block` we provide will be the only set of IP addresses that will be able to reach our database, cluster, and any other resources we have set up.

- **VPC**: The `vpc.tf` file will create a set of networks, connect them with routing, and also use our security group to limit access to them. These networks will be used by the Kubernetes cluster to deploy Pods, by the database, and by the application. The VPC will be installed across two AZs to improve our deployed resource resiliency by being installed in different data centers.

I have included the URL as a comment above each resource and module that is being used so you can visit and learn more about the resources being installed or learn about what other configuration options are available.

Next up is to deploy all of this infrastructure into AWS.

## Deploying the infrastructure

To create our deployment plan using the variable provided in the `terraform.tfvars` file, or when prompted, and to deploy it into AWS, we run the following command:

```
make deploy
```

This command will execute the `plan` and `apply` Terraform commands. These will be followed up with a command to fetch the cluster configuration so that we can connect to it with K9s. The plan that Terraform creates will contain approximately 87 resources. During the `apply` stage, Terraform will make use of the plan and will immediately begin the process of creating, configuring, connecting, and verifying each resource. Terraform will do its best to create resources concurrently when it can, but this process will take some time to complete – around 15 to 20 minutes.

> **Usage costs warning**
>
> Running these Terraform commands will create AWS resources that are not covered by any free tiers. You will begin incurring usage costs from the moment you execute the `make deploy` command. You will continue to be charged until you destroy the infrastructure with `make destroy`. Running this demo for a few hours will cost roughly $2 to 5 depending on the region that it is run in.

As it creates resources, Terraform will output logs of what is happening so that you are not left in the dark. You can also see some progress if you go into the AWS console and view the various services, such as EKS, RDS, and ECS.

If the process is interrupted or something times out, Terraform will end with an error. If it does, you can rerun the `make deploy` command to get things back on track in most cases.

When it is done, it will output any outputs we have defined to the screen as long as they are not marked sensitive. Some of these outputs will be used in our second phase of deploying the application.

## Viewing the Kubernetes environment

At this point, the infrastructure is completely set up. We can go into the AWS console to view various things, but if we try to view the Kubernetes cluster in EKS, it may say our user does not have permission to view the components. This is expected because we only gave the user we created permissions and not our main AWS account user. To view the cluster components, we will need to run the following command to bring up the K9s UI:

```
k9s
```

It might take a moment to load up completely but after it is done loading, we should see something like this:

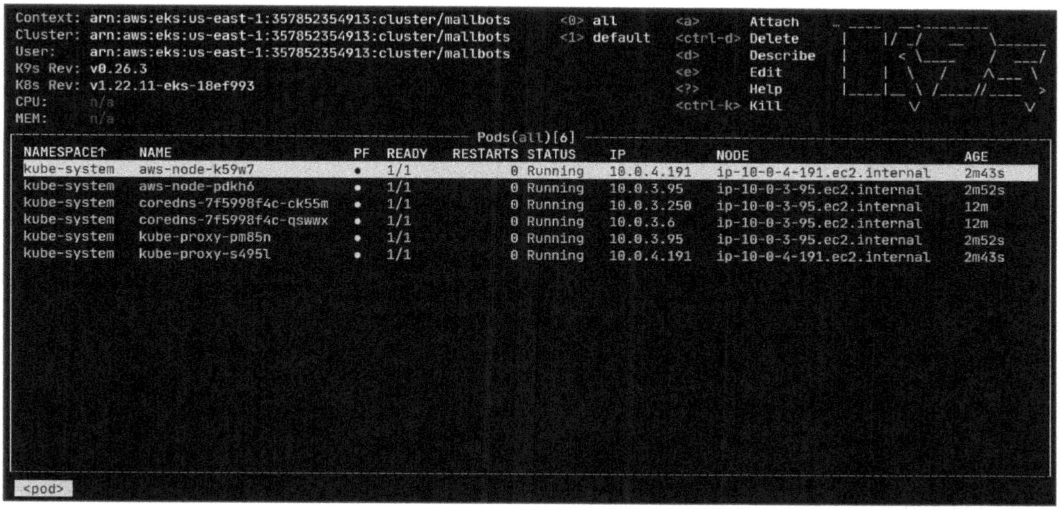

Figure 11.6 – The K9s terminal application showing the running Pods

To navigate around the components, you start the command with a colon and then the type of components you would want to view – for example, typing `:deployments` and hitting *Enter* will show the list of deployments in the cluster and `:services` will show the running services.

To exit K9s, type `:quit` and then hit *Enter*.

If you are familiar with `kubectl` or would prefer to work from the CLI instead, then to view the list of deployments, you can use the following command:

```
kubectl get deployment -n kube-system
```

This will display a short list of deployments in the `kube-system` namespace. Likewise, we can view the list of services using this command:

```
kubectl get services -n kube-system
```

This would display a short list of services.

Using either K9s or `kubectl`, we should see some load balancer resources installed with `load-balancer-aws-load-balancer-controller` in the list of deployments and `aws-load-balancer-webhook-service` in the list of services. Seeing these means that we know our infrastructure is ready.

Next, we need to deploy our application in the infrastructure that we have just deployed.

# Deploying the application to AWS with Terraform

To deploy the application, we will need to switch to the `/deployment/application` directory.

Similar to what we did for the infrastructure, we will prepare Terraform by installing the libraries that deploying the application will require by running the following command:

```
make ready
```

## Getting to know the application resources to be deployed

As we did for the infrastructure, we have broken up the resources we will be deploying into multiple files.

### Database setup

For the database, we will initialize the shared triggers and that action can be found in the `database.tf` file.

### Kubernetes setup

In Kubernetes, components can be organized into namespaces. This can help when you have multiple applications, when you have multiple users and want to restrict access, or when you are using the cluster for multiple purposes. Our application will be deployed into the `mallbots` namespace. In K9s, we can filter what we see by namespace to make it easier to locate just our application components.

As with our local development environment, the services will be using environment variables. Most of those variables are the same for each service and in Kubernetes, we can create `ConfigMaps` for data that we want to share. A config map is created with the common environmental variables, such as `ENVIRONMENT`, `WEB_PORT`, and `NATS_URL`. We will pass this config map into each microservices deployment resource.

Lastly, in the `kubernetes.tf` file, we define an ingress on the ALB, for the Swagger UI. Just as with our local experience, we will be able to visit a single URL to access all of the microservices and Swagger.

### NATS setup

In the `nats.tf` file, we create a deployment for NATS using the same container we used in the `docker compose` environment. A persistent volume claim, a little bit like the Docker volumes, is also set up for NATS to record its data. This way, if the deployment was restarted, we would not lose any messages. The NATS deployment is made available using a service component. A service defines how a deployment may be accessed.

### Microservices setup

Each microservice is kept in its own file using a filename pattern such as `svc_<service>.tf`.

Instead of using a static database password as we do in our local environment, each service uses a randomly generated password. These passwords are generated each time we plan and deploy the application freshly. Updating the application and redeploying will reuse the password from the Terraform state data. The random passwords are used within the initialized service database resource.

Kubernetes config maps are not good places to put secrets such as database passwords. They are not stored with any encryption so it is possible the data could be seen. For things such as passwords, we have secrets that do use encryption and are less likely to be seen or understood if they are leaked. For the `PG_CONN` environment variable, we create a secret and store each microservice separately.

As with NATS, each microservice has a deployment and a service component. Unlike NATS, most services also have an ingress setup so that they are also available at the exposed address provided by the ALB. Services such as `cosec` and notifications do not have any ingress defined because they do not expose any APIs.

## Deploying the application

To deploy the application in the waiting infrastructure, we run the following command:

```
make deploy
```

The application deployment consists of approximately 57 resources. This deployment will not take as long as the infrastructure deployment but will still clock in at around 5 to 10 minutes.

If you have K9s open, you can watch as the deployments come online and you can see the ingresses being added, using the `:deployments` and `:ingress` commands, respectively.

To view the list of deployments using `kubectl`, you would use the following command:

```
kubectl get deployment -n mallbots
```

This will display the following list, with different values in the AGE column, after all the deployments are done:

```
NAME            READY   UP-TO-DATE   AVAILABLE   AGE
baskets         1/1     1            1           18m
cosec           1/1     1            1           18m
customers       1/1     1            1           18m
depot           1/1     1            1           18m
nats            1/1     1            1           19m
notifications   1/1     1            1           18m
ordering        1/1     1            1           18m
payments        1/1     1            1           18m
search          1/1     1            1           18m
stores          1/1     1            1           18m
```

Please note that we are viewing the deployments in the `mallbots` namespace and not the `kube-system` namespace this time.

When the deployment has been completed, Terraform will output the address you can find in the Swagger UI. We are not deploying the application to any particular domain, so this address will be generated. If you missed the address, you could retrieve it using this command:

```
terraform output swagger_url
```

Opening this Swagger UI will be exactly the same as the experience we have locally. That is why IaC and repeatable deployments are so popular.

The application and infrastructure will only be accessible to your IP address but leaving it running will continue to cost you money and thankfully, solving this is also made easy using Terraform.

## Tearing down the application and infrastructure

Running this application in Kubernetes and using the infrastructure resources will continue to rack up costs by the hour for you so when you are done with the MallBots application, you should tear it all down. When Terraform makes changes to an environment, it keeps a state file – in our case, kept locally – so that it can minimize the changes it will need to make when the Terraform files are changed and new `plan` and `apply` commands are run. The state is also used to locate the resources that need to be destroyed when we are done with them.

Start with the application deployment first. Go into the `/deployment/application` directory and run the following command:

```
make destroy
```

As with the deployment process, this can take some time to complete. When it does complete, we can run the same command from the `/deployment/infrastructure` directory.

After the second command completes, your AWS account should be back to how it was before we started this journey. You can verify by signing into your AWS account on the AWS console and by visiting RDS to make sure there are no database instances, ECR to verify that there are no repositories, and EKS to see that the cluster has been completely removed. Anything you find across AWS you can view tags for; if you see the `Application: MallBots` tag, then it was something left behind. I ran and reran the deployment and tear-down steps over a dozen times and Terraform always did an excellent job restoring my account to how it was.

## Summary

In this chapter, we converted the modular monolith application into a microservices application. We modified the modules in such a way that we could continue to run the application as a monolith or with microservices. This is not exactly a goal most teams have but we could do it, so we did. A real application would likely begin to diverge and maybe pick up new microservices that are written in different languages, which would make keeping the monolith around an unlikely outcome.

We also set up our environment to deploy our application into the cloud using either a containerized approach or installing the necessary tools directly onto our system. We used these tools to stand up the infrastructure that our application needed to be run on top of first. Then, as a second step, we deployed the application itself to AWS. The experiences between the locally running application as a monolith, as microservices, and as a cloud deployment remained exactly the same.

In the next chapter, we will be learning how to monitor the performance of our application and to track requests as they flow through it using causation and correlation identifiers.

# 12

# Monitoring and Observability

In this final chapter, we will cover how to monitor the performance and health of the services and the application as a whole. The most common approach to monitoring is to use logging to record information and errors. Following logging, the second most common approach is to record metrics such as CPU usage, request latency, and more. We will be looking into these forms of monitoring and will also take a look at an additional form of monitoring, known as distributed tracing.

In this chapter, we will also introduce OpenTelemetry and learn about its goals, and how it works. We will then add it to our application to record each request as it works its way through the application.

We will end by looking at the tools that are used to consume the data produced by our monitoring additions – that is, Jaeger, Prometheus, and Grafana.

In this chapter, we are going to cover the following main topics:

- What are monitoring and observability?

- Instrumenting the application with OpenTelemetry and Prometheus

- Viewing the monitoring data

## Technical requirements

You will need to install or have installed the following software to run the application or try the examples provided:

- The Go programming language version 1.18+

- Docker version 20+

- Docker Compose version 2+

The code for this chapter can be found at `https://github.com/PacktPublishing/Event-Driven-Architecture-in-Golang/tree/main/Chapter12`.

# What are monitoring and observability?

Most deployments perform monitoring using logging and metrics. This allows an organization to track the application's performance, usage, and health. It is also used to detect failure states in an application. Monitoring is about reacting to the analysis of the data that is being collected:

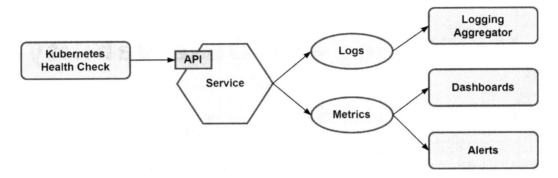

Figure 12.1 – Basic monitoring of a service

Some examples of monitoring include the following:

- Kubernetes checking whether a container is still running or responding by performing a health check
- Tracking the query performance when swapping out one database for another
- Automatically scaling services based on the CPU and memory usage
- Sending alerts when the error rate of an endpoint exceeds a certain threshold

The data that is produced from your monitoring efforts is fed into dashboards so that basic questions can be answered. The data is also used to configure alerts so that when a problem is developing, staff can be notified to take the appropriate action.

Monitoring works with predetermined logs and metrics. Its weakness is dealing with the unexpected. If we know ahead of time that a process has the potential to consume large amounts of CPU, then we can include that in our monitoring; otherwise, it will be a blind spot. Putting this another way, if we can predict that we will have problems in a certain part of the application, then we can instrument it so that it can be monitored.

The data that is collected from the various monitoring efforts is often used in isolation. As a result, it can sometimes be difficult to correlate an event across different sets of data. Keeping the data isolated is not done by choice; the tools themselves are typically not designed to interact with other tools or other forms of data.

The purpose of monitoring is to answer "*What happened?*" and "*Why did it happen?*". However, when we need to correlate the data ourselves across different tools, it is not always easy. For example, let's say that a team receives an alert about a service experiencing rising CPU usage. To determine the root cause, the team could look at related dashboards to determine a timeframe, and then search the logs to locate any errors during the timeframe reported in the dashboards. If the team fails to locate any errors, a new search to look over all the logs to spot a trend would be necessary. This is a typical approach many teams take, and it can be a workable solution for most applications. Distributed applications only make the problem of locating the root cause of an issue more difficult. With a distributed system, requests travel through multiple services and use a variety of communication methods.

As the application grows in complexity, so does the need to monitor for more things. Making accurate predictions about all of the places that will be problematic can be extremely difficult. You need instrumentation that will be able to answer the questions about the *unknown unknowns* or to provide answers without asking explicit questions. This is where observability and distributed tracing enter the picture.

## The three pillars of observability

Observability is made up of *three pillars*. We discussed the first two – logs and metrics – in the previous section; in this section, we will be covering the third: traces. Traces are recordings of a request as it moves through the application.

Together with logs and metrics, traces give you a complete picture of the state of the application:

- Logs tell you why your application is in a given state
- Metrics tell you how long your application has been in a given state
- Traces tell you what is impacted by being in a given state

A trace may begin with the client being at the first entry point in the application or even somewhere in between. The trace will be given some kind of identifier that is passed along as the request that it tracks makes its way through the application.

## How tracing works

We will work with an example where the trace starts as it enters the application's backend. When a brand-new request comes in, an identifier is generated for it; for example, **abcd**. At the same time, correlation and causation identifiers are also assigned the same value:

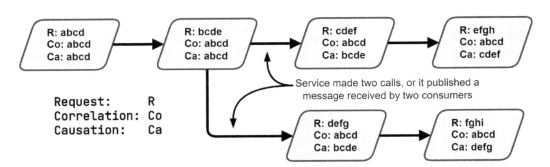

Figure 12.2 – Tracing with request, correlation, and causation identifiers

The purpose of a correlation identifier is to correlate all requests back to a single originating request. A causation identifier is used to point back from a follow-up request to the request that came before it.

As the request makes its way through the application, the correlation identifier never changes. The causation identifier will always point back to the call that preceded it. Requests into the application can fork. Here, we follow the same rules that have already been laid out; no new rules are required to handle branches that can occur during a request.

These identifiers can then be logged with other log messages. If you are building your tracing implementation manually, then this is how you might record how a request flows through the application. You will not be able to construct a span, a representation of a call, or another unit of work with log messages alone.

Tools such as **Jaeger** can visualize a trace, giving you an entirely new view of your application that you can't see from metrics or the logs themselves.

In a visualization of a trace, you can see the different spans along the *y*-axis, which could be different processes that were run in different components. Along the *x*-axis, you can see the element of time so that you can get a sense of how long it took to log those processes:

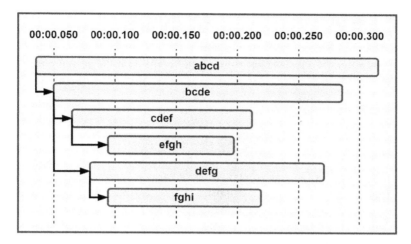

Figure 12.3 – Visualization of a request traced through the application broken into multiple spans

You could develop your own tracing implementation, but I would suggest otherwise. There is a lot more than simply being able to visualize an icicle diagram, or upside-down flame graph, of the different spans that make up the trace. Traces cannot be created from log messages, so you would be developing an entirely new instrumentation method for your application. Traces are also very information-rich and can be annotated with bits of information and even record errors that have occurred at specific points.

Thankfully, you do not need to start from scratch to instrument your application. The **OpenTelemetry** project (`https://opentelemetry.io`) exists for this purpose, which has the goal of merging the instrumentation for logging, metrics, and tracing into a single unified API.

## Instrumenting the application with OpenTelemetry and Prometheus

Our application has already been set up with a logger, but we need traces and metrics to achieve observability. The OpenTelemetry project aims to support all three (logging, traces, and metrics) in the **Go SDK** but at the time of writing this book and version v1.10, only tracing is stable. So, we will leave our logger in place and interact directly with Prometheus for our metrics. We will begin with OpenTelemetry and distributed tracing.

## Adding distributed tracing to the application

Getting started with OpenTelemetry is very easy; first, we will need to create a connection to the collector. In our application, we will have one running and available at the default port. The monolith or microservices will use the following environment variables to configure themselves:

```
OTEL_SERVICE_NAME: mallbots
OTEL_EXPORTER_OTLP_ENDPOINT: http://collector:4317
```

The OpenTelemetry SDK we will use will look for specific variables in the environment that all begin with the OTEL prefix, short for OpenTelemetry. The two variables shown in the preceding snippet are the minimum we will need to run our demonstration. OTEL_SERVICE_NAME should be set to a unique name for the application or component. Here, we are setting it to mallbots for the monolith application. For the services, we will use their package names. The SDK defaults to looking for the collector on localhost. This could work, but we have set it to the hostname we have given it in the Docker Compose environment. We will be communicating with the collector using the **OpenTelemetry Protocol** (**OLTP**) and will set OTEL_EXPORTER_OTLP_ENDPOINT to our collector host and the OLTP port.

An OpenTelemetry collector is a vendor-agnostic service that provides instrumentation data collection, processing, and exporting functionality. A single collector can replace the need to run, configure, and connect to multiple agents to instrument your application.

> **The demonstration is local only**
>
> There is no advantage to running the demonstration of the application instrumentation in AWS, so this demonstration is expected to be run locally in your Docker Compose environment.

Now, we can update the /internal/system code to initialize the connection to the collector:

```
func (s *System) initOpenTelemetry() error {
    exporter, err := otlptracegrpc.New(
        context.Background(),
    )
    if err != nil { return err }
    s.tp = sdktrace.NewTracerProvider(
        sdktrace.WithBatcher(exporter),
    )
    otel.SetTracerProvider(s.tp)
    otel.SetTextMapPropagator(
        propagation.NewCompositeTextMapPropagator(
```

```
            propagation.TraceContext{},
            propagation.Baggage{},
        ),
    )
    return nil
}
```

The `initOpenTelemetry()` method will set up a gRPC connection to the collector; we do not need to provide any host or address information because we have already set that in the environment. Then, we need to set up the tracer so that it sends the trace data to the collector in batches. This helps improve performance and should be used in most cases.

The tracer provider, `s.tp`, is then set as the default. Now, anywhere in the application that we need to interact with the tracer provider, we can simply call it up and do not need to pass a reference into our structures or include a value for it in the context. Not having to do either of those things makes it very easy to adopt the library into your application.

The function finishes by setting the default for how trace data should be propagated. Both the normal trace context data and any optional baggage, such as additional metadata, will be read and saved as maps of strings.

The gRPC server and the message handlers will also need to be updated with new middleware so that the new spans are created automatically.

For the gRPC server, the OpenTelemetry Go library provides client and server interceptors that we can quickly add. For the server, we must add the following to the gRPC server initializer:

```
s.rpc = grpc.NewServer(
    grpc.UnaryInterceptor(
        otelgrpc.UnaryServerInterceptor(),
    ),
    // If there are streaming endpoints also add:
    // grpc.StreamInterceptor(
    //     otelgrpc.StreamServerInterceptor(),
    // ),
)
```

Adding the interceptor for clients is also a straightforward affair. The interceptors for clients are added as a `Dial` option:

```
func Dial(ctx context.Context, endpoint string) (
    conn *grpc.ClientConn, err error,
```

```
) {
    return grpc.DialContext(ctx, endpoint,
        grpc.WithTransportCredentials(
            insecure.NewCredentials(),
        ),
        grpc.WithUnaryInterceptor(
            otelgrpc.UnaryClientInterceptor(),
        ),
        // If there are streaming endpoints also add:
        // grpc.WithStreamInterceptor(
        //     otelgrpc.StreamClientInterceptor(),
        // ),
    )
}
```

For our message publishers and subscribers, the OpenTelemetry library does not have any ready-made middleware for our custom code, but creating a couple of new middleware is an easy task.

In a new package, /internal/amotel, which has been named as such to signify an instrumentation relationship with the /internal/am package, we have the OtelMessageContextInjector() and OtelMessageContextExtractor() functions. We use the injector for all our outgoing messages, so every MessagePublisher constructor call will be updated to receive it as a new middleware:

```
am.NewMessagePublisher(
    stream,
    amotel.OtelMessageContextInjector(),
    tm.OutboxPublisher(outboxStore),
)
```

We need to be careful regarding the order in which we apply the middleware for the outbox and the new instrumentation. If we put them in the wrong order, then the messages we store in the outbox will not be modified with the correct metadata.

We use the extractor in the MessageSubscriber constructor calls:

```
am.NewMessageSubscriber(
    stream,
    amotel.OtelMessageContextExtractor(),
)
```

This time, there is no issue with any existing middleware and there are no ordering concerns. Now, so long as we have covered every constructor for the publishers and subscribers, our application will output span data for each traced request.

Previously, I mentioned that traces can contain more data; now, we can see what some of that additional data might be. Upon opening the `/baskets/internal/grpc/server.go` file, we will find that the server calls have been updated with new instrumentation. For example, take a look at the following excerpt from `CheckoutBasket()`:

```
span := trace.SpanFromContext(ctx)

span.SetAttributes(
    attribute.String("BasketID", request.GetId()),
    attribute.String("PaymentID", request.GetPaymentId()),
)
```

In the first line, we get the current span; if one doesn't exist, the library will return a **no-op** – a no operation – span to us so that our code does not break and then the next lines annotate it with some values that are important to this gRPC request. These attributes are not being put into the span so that they can be recalled later, like how placing values into a context works. Attributes are information that will be sent to the trace collector and can be used to diagnose requests from tools such as Jaeger.

You can interact with the existing span using `trace.SpanFromContext(ctx)`. You can also create new spans for processes that should be given their own spans, such as intense processing tasks. To create a new span, you can use code similar to the following:

```
ctx, span := otel.GetTracerProvider().
    Tracer("pkg_name").
    Start(ctx, "span_name")
```

This will grab the default tracer provider, then create a new tracer with whatever name you want to give it. But the best practice is to use the fully qualified package name. Then, a new span will be started, using any span from the context as the parent, and any name you wish to use.

Unless you know you need to create a new span, it is best to work with the existing span from the context.

Traces can also be annotated with events. Here, events are annotations that also have time components. This is very similar to logging, but the data is encapsulated entirely within the trace data. These too can be visualized in the graphs and diagrams produced by the trace tools. The event is visualized as either a line or dot on the span it was recorded to. Using events adds another dimension to the data that makes the flow of time more apparent:

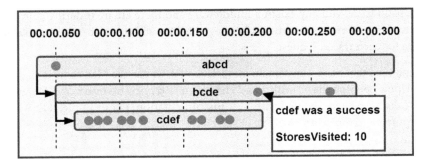

Figure 12.4 – Spans annotated with events

The domain event handlers in each module will record additional information about the events they handled and the amount of time that it took. The following excerpt is from the `/baskets/internal/handlers/domain_events.go` file:

```
span := trace.SpanFromContext(ctx)
defer func(started time.Time) {
    if err != nil {
        span.AddEvent(
            "Error encountered handling domain event",
            trace.WithAttributes(
                errorsotel.ErrAttrs(err)...,
            ),
        )
    }
    span.AddEvent("Handled domain event",
        trace.WithAttributes(
            attribute.Int64(
                "TookMS",
                time.Since(started).Milliseconds(),
            ),
        ))
} (time.Now())
span.AddEvent("Handling domain event",
    trace.WithAttributes(
        attribute.String("Event", event.EventName()),
    ),
)
```

In the preceding snippet, we are adding events before and after the event is handled. If handling the domain event results in an error, then a third event is going to be added with information about the error itself. When these two or three events are displayed in the graphs, they are positioned proportionately to the entire trace when they occurred.

I could have also recorded the error directly to the span using `RecordError()`. Doing this would change the status of the span to reflect that an error was encountered. Likewise, I could also directly set the status of the span when an error existed with `SetStatus()`. I do not want to use either here because I only want to record the fact an error occurred. The middleware that is used for the gRPC server and on `MessageSubscriber` will take care of calling both of those functions if the error hasn't been handled already. Once you record an error to a span or set the status to the error level, you cannot undo it. So, it is best to let the code that created the span take care of doing both.

This is all the distributed tracing we will be adding in this chapter, but do experiment with updating a module or two to play around with creating new spans, adding attributes, and recording events.

To instrument the application with OpenTelemetry, we made the following updates:

- Created a default `TracerProvider` struct in the `internal/system` package, which is configured using environment variables using a new method named `initOpenTelemetry()`

- Added gRPC interceptors to the server and client dialers to propagate the trace context for gRPC requests

- Added middleware to `MessagePublisher` and `MessageSubscriber` to propagate the trace context for messages

- In each gRPC server, we annotated the spans with relevant request data

- The domain event handlers were updated to bookend the handling of the domain events by recording events in the span

Next, we will learn how to report metrics about the application to Prometheus.

## Adding metrics to the application

We will be using **Prometheus** (https://prometheus.io/) to instrument our application to report metrics. Prometheus is quick to set up and just as quick to use.

To begin, we need to set up an endpoint on the HTTP server so that Prometheus can fetch the metrics we will be publishing. Unlike OpenTelemetry, which uses a push model to send data to the collector, Prometheus uses a pull model and will need to be told where to look for metrics data.

To provide Prometheus an endpoint to fetch the data, we need to import the `promhttp` package and then add the handler it provides to the HTTP server. We must modify the `/internal/system/system.go` file to add the endpoint:

```
import (
    "github.com/prometheus/client_golang/prometheus/promhttp"
)
// ... much further down
func (s *System) initMux() {
    s.mux = chi.NewMux()
    s.mux.Use(middleware.Heartbeat("/liveness"))
    s.mux.Method("GET", "/metrics", promhttp.Handler())
}
```

Prometheus expects to find metrics at the `/metrics` path by default, but that can be changed when you configure Prometheus to fetch the data.

The Go client for Prometheus automatically sets up a bunch of metrics for our application. Hitting that endpoint will display a dizzying list of metrics that were set up for us for free. We can also set up custom metrics for our application; to demonstrate, we will start with the messaging system.

When publishing a message, we must use a counter to record a total count and a count for that specific message:

```
counter := promauto.NewCounterVec(
    prometheus.CounterOpts{
        Namespace: serviceName,
        Name:      "sent_messages_count",
        Help:      fmt.Sprintf(
            "The number of messages sent by %s",
            serviceName,
        ),
    },
    []string{"message"},
)
```

This is setting up a monotonically increasing counter that is broken up into partitions using a message value. The message value will be whatever is returned by calling `MessageName()` on the outgoing message. The service name is used as a namespace to avoid collisions when we are reporting metrics from the monolith. The namespace will be prefixed to the counter name, changing its name to something like `baskets_sent_messages_count`.

We are also using the `promauto` package to register these new metrics automatically with the default registry. If we were not using the `promauto` package and were using the `prometheus` package instead, we would need to include the following line to register the counter:

```
prometheus.MustRegister(counter)
```

To record both the total count and the individual message count, we can use the following two lines:

```
counter.WithLabelValues("all").Inc()
counter.WithLabelValues(msg.MessageName()).Inc()
```

Each time a message is published, we increment two partitions – the `all` partition and the message-specific partition.

The values kept in the counter will be lost when the service is restarted, and that is fine in most cases. Counter metrics are typically going to be watched for trends such as increasing too quickly, staying level over time, and so forth. The actual value of the counter rarely comes into play.

On the receiving side, we will use a similar counter to record how many messages have come in:

```
counter := promauto.NewCounterVec(
    prometheus.CounterOpts{
        Namespace: serviceName,
        Name:      "received_messages_count",
        Help:      fmt.Sprintf(
            "The number of messages received by %s",
            serviceName,
        ),
    },
    []string{"message", "handled"},
)
```

This time, the counter has a second label called `handled`, which will be used to further split the count into successfully handled messages and the ones that produced an error. We are also interested in how long it takes to handle a message, so we will use another type of metric: a histogram.

Histograms are used to track length-like values such as request duration or message size. They are configured with buckets that will store the counts. We will use one to record the time it takes to handle each incoming message:

```
histogram := promauto.NewHistogramVec(
    prometheus.HistogramOpts{
        Namespace: serviceName,
```

```
    Name:        "received_messages_latency_seconds",
    Buckets:     []float64{
        0.01, 0.025, 0.05, 0.1,
        0.25, 0.5, 1, 2.5, 5,
    },
},
[]string{"message", "handled"},
)
```

Like the counter, we will use two labels to partition the histogram. The Buckets field is optional, and Prometheus provides a default bucket setup very similar to what's shown in the preceding code example.

To record all of the metrics for the incoming messages, we will use the following code. This will record four metrics when handling a message:

```
handled := strconv.FormatBool(err == nil)
counter.WithLabelValues("all", handled).Inc()
counter.WithLabelValues(
    msg.MessageName(), handled,
).Inc()
histogram.WithLabelValues(
    "all", handled,
).Observe(time.Since(started).Seconds())
histogram.WithLabelValues(
    msg.MessageName(), handled,
).Observe(time.Since(started).Seconds())
```

For each metric, we record the all partition and the specific message partition. To determine whether the message was handled properly, we check whether the err value is nil. This will record the metrics on a lot of partitions, which can be useful in setting up detailed dashboards.

The metrics are recorded using middleware that lives in the /internal/amprom package. Using this middleware is going to be the same as using the OpenTelemetry middleware we created. For the publisher, we can add it before the outbox middleware:

```
am.NewMessagePublisher(
    stream,
    amotel.OtelMessageContextInjector(),
    amprom.SentMessagesCounter("baskets"),
    tm.OutboxPublisher(outboxStore),
)
```

Then, we can use the same ID we used for the `NewMessageSubscriber` constructor by adding it either before or after the OpenTelemetry middleware:

```
am.NewMessageSubscriber(
    stream,
    amotel.OtelMessageContextExtractor(),
    amprom.ReceivedMessagesCounter("baskets"),
)
```

We will be able to create detailed dashboards showing the number of messages being used in our application and how long it takes our application to process each one.

Speaking of dashboards, they are not only used by the engineers working on the application but also by people from other departments. It is common to expose metrics about how much product is being produced, or how many customers have registered. We can add those kinds of metrics as well.

In the composition root for the **Customers** module, we can add a counter for `customers_registered_count`:

```
customersRegistered := promauto.NewCounter(
    prometheus.CounterOpts{
        Name: "customers_registered_count",
    },
)
```

There's no need for a namespace or partitions this time; we can use a simple counter. We want to use this counter to count every successful registration that is made. We could pass the counter into the application, then increment the counter if there was no error being returned by the `RegisterCustomer()` method by checking the results with a deferred function. This would not be my first choice on how to go about this. The MallBots application is a relatively simple application and the `Application` struct in the real application may already be dealing with a lot of dependencies. My preference is to create a wrapper for the `Application` struct that will be used for this counter and any other metric we want to add. This keeps the concerns separated and keeps the existing `Application` tests unchanged. It also means we can test the wrapper in isolation.

The wrapper will only intercept the `RegisterCustomer()` method, letting all of the other methods pass through unaffected:

```
type instrumentedApp struct {
    App
    customersRegistered prometheus.Counter
}
```

```
func NewInstrumentedApp(
    app, customersRegistered prometheus.Counter,
) App {
    return instrumentedApp{
        App:                 app,
        customersRegistered: customersRegistered,
    }
}

func (a instrumentedApp) RegisterCustomer(
    ctx context.Context, register RegisterCustomer,
) error {
    err := a.App.RegisterCustomer(ctx, register)
    if err != nil { return err }
    a.customersRegistered.Inc()
    return nil
}
```

To use this instrumented application, we need to wrap the application instance in the composition root:

```
application.NewInstrumentedApp(
    application.New(
        customersRepo,
        domainDispatcher,
    ),
    customersRegistered,
)
```

Other modules can be updated to record metrics such as counting the number of baskets started by the users or counting the number of new products made available by the stores.

Let's recap what we did to add Prometheus metrics to the application:

- An endpoint was added to the HTTP service so that Prometheus can retrieve our metrics
- Middleware was added to add metrics for the published and received messages
- Middleware was included in the constructors for the MessagePublisher and MessageSubscriber interfaces

- Additional application counters were created, such as the registered customer counter

- An application wrapper was used to instrument the application without modifying it

In this section, we added distributed tracing and metrics to our application. This covers all three pillars of observability since the application already had logging. Everything that was added should have no measurable impact on the application; if it does, we will now be able to monitor it.

In the next section, we will learn about the tools we can use to view the data that's now being reported about the application.

## Viewing the monitoring data

The application will now be producing a lot of data; to view this data, we need to collect it or, in the case of Prometheus, retrieve it.

The Docker Compose environment was updated with four new services, as follows:

- The OpenTelemetry collector, which will collect trace and span data

- Jaeger to render the traces

- Prometheus to collect and display metrics data

- Grafana to render dashboards based on the metrics data

The **OpenTelemetry** collector will also provide **Prometheus** metrics about the traces and spans it collects:

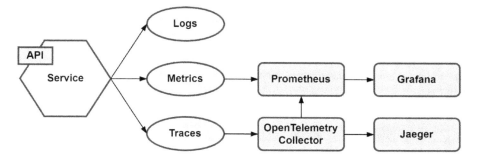

Figure 12.5 – The additional monitoring services

We have already configured the modules to connect with the collector so that is ready to go. For Prometheus, we still need to configure it to retrieve the metrics from each microservice. The configuration file, `/docker/prometheus/prometheus-config.yml`, will need to be updated so that it contains a job for each microservice we want to scrape. For the **Shopping Baskets** microservice, we must add the following under the `scrape_configs` heading:

```
- job_name: baskets
  scrape_interval: 10s
  static_configs:
    - targets:
      - 'baskets:8080'
```

There are a lot more options we could set here, but these are all we will need for now.

At this point, we can start up the Docker Compose environment, then use the **Swagger UI** to make some requests. However, making individual requests with the Swagger UI could take some time; we need to build up enough data to give us some idea of what collecting data from an active application might look like.

Instead, we can use a small application that can be found under `/cmd/busywork` to simulate several users making requests to perform several different activities. The application is nothing fancy and you are encouraged to modify it to simulate whatever interactions you like.

With the MallBots application already running locally with Docker Compose, start the busywork application by running the following:

```
cd cmd/busywork
go run .
```

Five clients will be started up and will begin making requests. You can increase the number of clients by passing in the `-clients=n` flag, with an upper limit of 25. To end the busywork application, use *Ctrl + C* or *Cmd + C*; this will kill the process.

Now, we can look at some of the data that is being produced, starting with **Jaeger**. Open `http://localhost:8081` in your browser to open Jaeger. You should see a UI like this:

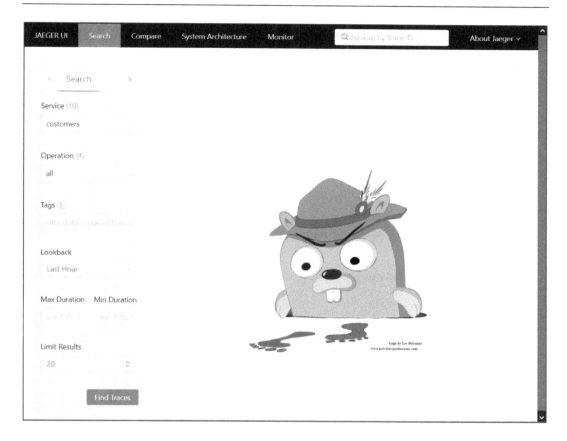

Figure 12.6 – The Jaeger UI

Toward the left, under **Service**, select the **customers** service and click the **Find Traces** button. Doing this will show several traces in a timeline view and as a list. In the timeline, the size of the circle signifies the size of the trace. The larger the circle, the more spans that were involved. Also, the height of the circles signifies the duration of the trace. This is an example of a search for traces that involve the **customers** service; your graph will be different because the busywork clients will be randomly interacting with the MallBots application:

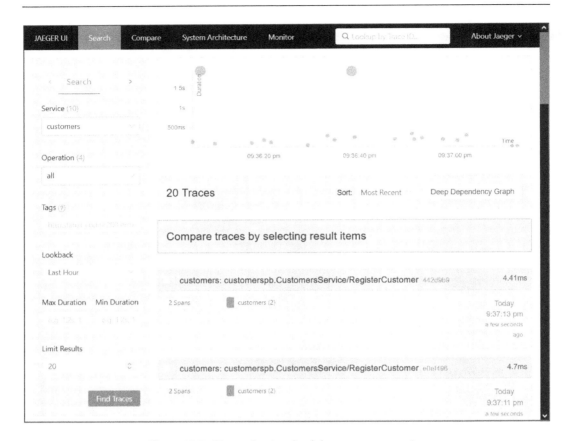

Figure 12.7 – Traces that involved the customers service

If you do not have any of the larger trace circles, as shown in the preceding figure, wait for a moment and perform a new search; eventually, one will appear. These larger circles are from the create order saga execution coordinator. If you click on one, it will open up the trace details screen for that trace. From the details screen, we can see how the services all worked together to accomplish the task of creating a new order:

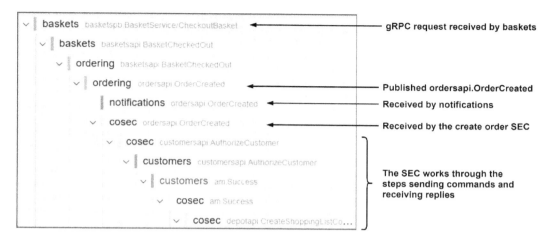

Figure 12.8 – A portion of the create order process shown in Jaeger

Clicking on one of the rows in the graph will provide you with additional details. If we click on the first row for **baskets basketspb.BasketService/CheckoutBasket**, we will be able to see the additional data we recorded to the span using the gRPC service's `CheckoutBasket()` method. Under **Tags**, we will find the `BasketID` and `PaymentID` properties, which were used for this request. Under **Logs**, we will find the events that were recorded to the span ordered by time, with all times relative to the start of the trace.

Remember when we added the bookend events to the handling of the domain events? If you compare the log timestamps for the two events with the timestamps of the next child span, you will see that the second log correctly shows it occurred after the child span had been completed.

A lot of data recorded is with each trace and that is its major downside. Recording a trace can be very demanding on the disk to store them, the CPU to process them, and the network to collect them. To lessen this resource demand, traces are sampled and only some are saved. Deciding to save a trace is either head-based, during the initial parent span creation, or tail-based, where the decision can be made by a child span at any time. OpenTelemetry only supports head-based decision-making. The upside is that it is easier to implement and work with, but the downside is that it drops traces that include errors that might be worth checking out. One day, OpenTelemetry might offer tail-based decision-making, but until it does, you should continue to use logging to capture important errors.

We also have the metrics to check out in Prometheus. Opening `http://localhost:9090` in your browser will present you with the following UI:

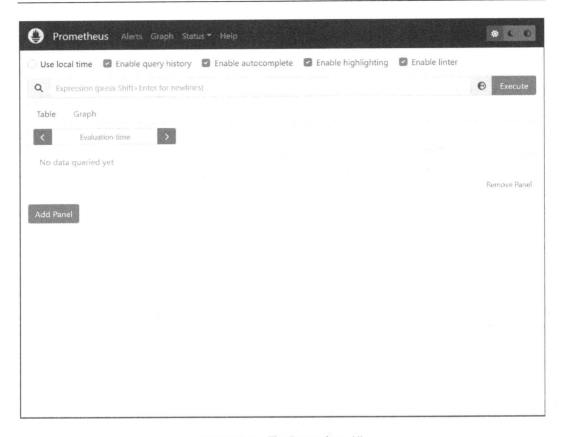

Figure 12.9 – The Prometheus UI

Performing a search for cosec_received_messages_count will return results similar to this:

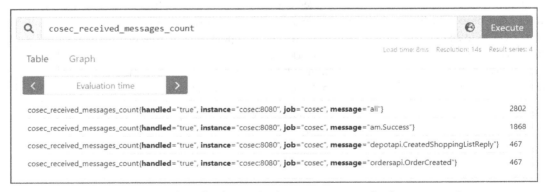

Figure 12.10 – Searching for the received messages counts for the cosec service

You could also try searching for `go_gc_duration_seconds` to see the garbage collector metrics for each microservice or any other metric you can think of. Like the traces, we are dealing with a very large amount of data – not as much as with the traces, but certainly a large number of metrics.

Searching for metrics in Prometheus and viewing the raw data is not very compelling. That is why we also have Grafana running. Opening `https://localhost:3000/` and then browsing for dashboards will show the two dashboards that are installed under the `MallBots` folder. The **Application** dashboard will display some panels that will give you insights into how active the application is, and will display several panels showing the rates of incoming and outgoing messages for a few services:

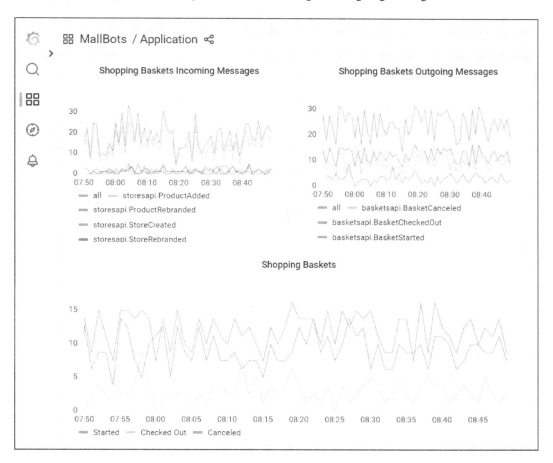

Figure 12.11 – The MallBots Application dashboard

How much activity you see in the dashboard will depend on how many clients you have running in the busywork application and the random interactions that the clients are performing.

The other dashboard that is available is the OpenTelemetry **Collector** dashboard, which will provide some details about how much work the collector is doing.

With a small to moderate amount of work, we added a massive amount of instrumented data to our application that gives incredible insight into the inner workings of the application.

## Summary

In this final chapter, we learned about monitoring and observability. We were introduced to the OpenTelemetry library and learned about its goals of making applications observable easier. We also learned about distributed tracing and how it is one of the three pillars of observability.

Later, we added both distributed tracing and metrics to the application using OpenTelemetry and Prometheus. With a little work, both forms of instrumentation were added to the application. To demonstrate this new instrumentation, we made use of a small application to simulate users making requests while we were free to view the recorded data in either Jaeger or Prometheus.

This chapter concludes the adventure we started, which involved taking a synchronous application and refactoring it to turn it into a fully asynchronous application that could be deployed to AWS and be completely observable.

# Index

## V

## W

Packt.com

Subscribe to our online digital library for full access to over 7,000 books and videos, as well as industry leading tools to help you plan your personal development and advance your career. For more information, please visit our website.

## Why subscribe?

- Spend less time learning and more time coding with practical eBooks and Videos from over 4,000 industry professionals

- Improve your learning with Skill Plans built especially for you

- Get a free eBook or video every month

- Fully searchable for easy access to vital information

- Copy and paste, print, and bookmark content

Did you know that Packt offers eBook versions of every book published, with PDF and ePub files available? You can upgrade to the eBook version at packt.com and as a print book customer, you are entitled to a discount on the eBook copy. Get in touch with us at customercare@packtpub.com for more details.

At www.packt.com, you can also read a collection of free technical articles, sign up for a range of free newsletters, and receive exclusive discounts and offers on Packt books and eBooks.

# Other Books You May Enjoy

If you enjoyed this book, you may be interested in these other books by Packt:

**Go for DevOps**

John Doak, David Justice

ISBN: 9781801818896

- Understand the basic structure of the Go language to begin your DevOps journey
- Interact with filesystems to read or stream data
- Communicate with remote services via REST and gRPC
- Explore writing tools that can be used in the DevOps environment
- Develop command-line operational software in Go
- Work with popular frameworks to deploy production software
- Create GitHub actions that streamline your CI/CD process
- Write a ChatOps application with Slack to simplify production visibility

**Mastering Go - Third Edition**

Mihalis Tsoukalos

ISBN: 9781801079310

- Use Go in production
- Write reliable, high-performance concurrent code
- Manipulate data structures including slices, arrays, maps, and pointers
- Develop reusable packages with reflection and interfaces
- Become familiar with generics for effective Go programming
- Create concurrent RESTful servers, and build gRPC clients and servers
- Define Go structures for working with JSON data

## Packt is searching for authors like you

If you're interested in becoming an author for Packt, please visit authors.packtpub.com and apply today. We have worked with thousands of developers and tech professionals, just like you, to help them share their insight with the global tech community. You can make a general application, apply for a specific hot topic that we are recruiting an author for, or submit your own idea.

## Share Your Thoughts

Now you've finished *Event-Driven Architecture in Golang*, we'd love to hear your thoughts! Scan the QR code below to go straight to the Amazon review page for this book and share your feedback or leave a review on the site that you purchased it from.

https://packt.link/r/1803238011

Your review is important to us and the tech community and will help us make sure we're delivering excellent quality content.

# Download a free PDF copy of this book

Thanks for purchasing this book!

Do you like to read on the go but are unable to carry your print books everywhere? Is your eBook purchase not compatible with the device of your choice?

Don't worry, now with every Packt book you get a DRM-free PDF version of that book at no cost.

Read anywhere, any place, on any device. Search, copy, and paste code from your favorite technical books directly into your application.

The perks don't stop there, you can get exclusive access to discounts, newsletters, and great free content in your inbox daily

Follow these simple steps to get the benefits:

1. Scan the QR code or visit the link below

https://packt.link/free-ebook/9781803238012

2. Submit your proof of purchase

3. That's it! We'll send your free PDF and other benefits to your email directly

www.ingramcontent.com/pod-product-compliance
Lightning Source LLC
Chambersburg PA
CBHW062046050326
40690CB00016B/2997